Hypnofacts 4

Trevor Eddolls

This book is dedicated to
Jill; Katy, Harry, and Freddy; and Jennifer and Jake

First published in 2016
By iTech-Ed Hypnotherapy
16 Brinkworth Close
Chippenham
Wilts SN14 0TL

Typeset by iTech-Ed Ltd

All rights reserved
© iTech-Ed Hypnotherapy, 2016

The right of Trevor Eddolls to be identified as the author of this work
has been asserted in accordance with Section 77 of
The Copyright, Designs, and Patents Act 1988
978-1-326-76167-7

Contents

Introduction	vii
Setting up a new practice	1
Solution-Focused Brief Therapy	3
Making changes	9
Hypnotherapy and exam nerves	19
Dealing with panic attacks	26
Fear of driving	29
What to do when a client is 'stuck'	33
Raising a client's self-esteem	40
The complete guide to overcoming any fear ever!	42
IBS and hypnotherapy	47
Hypnotherapy and fibromyagia	49
Laugh and the world laughs with you	51
Cortisone gets a bad press	53
Hardwiring	56
A funny tummy and my brain!	61
Diet and depression	65
The primitive brain gets a bad press	67
Unthinking thinking	70
Mindfulness and solution-focused hypnotherapy	75
Working with groups	81
Positive psychology	86
What can we learn from Sufism	90
About the author	96

Introduction

Like its predecessors, this book contains various articles for hypnotherapists covering practical issues such as helping clients with IBS and fibromyalgia, and information about how clients can make changes to their lives and overcome any fear ever. And there are more theoretical issues such as working with groups and some ideas from positive psychology, as well as what we can learn from the Sufis.

Again, the articles assume a model of the brain in which core activities (such telling the heart to beat) are handled by the 'reptilian' brain, more protective functions (such as fighting, fleeing, feeding, and reproductive behaviour) are handled by the primitive emotional brain, and higher functions (such as problem solving, maintaining attention, and controlling emotional impulses from the primitive brain) are handled by the intellectual brain. In terms of physical parts of the brain, these three areas more-or-less match up to the brain stem and cerebellum, the limbic system, and the cerebral cortex. It also assumes that the primitive emotional brain is very fast and the intellectual brain is much slower and tends to be used less.

In addition, the book assumes that the mind and body make up a single functioning system that is affected by the environment they're in.

And it assumes a solution-focused model for hypnotherapy – moving clients towards their desired goals rather than worrying about the problem itself and its origin.

Setting up a new practice

A look at some of the things you might want to think about when setting up a new practice.

Maybe you've just moved to a new town. Perhaps you've decided to make a move from working at home to working from a clinic. Or it might be that you've decided to work another couple of days as a hypnotherapist in a nearby town. The question is the same – how do I decide which clinic I want to work from?

Your first task is to find out what clinics there are in your town and the surrounding area. I say clinics, but it could be a room over a hairdressers or a room in a health food shop, or even a room at a gym – there are a lot of options available. For convenience, I'll call them all clinics – but you know what a wide range of establishments I have in mind. Once you've identified the options available, the next step is to go and have a look at them. What does the clinic look like from the outside? Is it somewhere that you would feel happy going for treatment? If not, cross it off your list and move on to the next clinic on your list. If you're based in a big city, where most people walk or use public transport, parking won't necessarily be an issue. But in most towns, your clients will mainly arrive by car and so somewhere to park could be an issue for them. Also, you may want to think about ease of access for your less mobile clients. Are there drop kerbs nearby? Is there ramped access and handrails to hold on to?

Having looked at the property from the outside, the next step is to go inside. Is there a welcoming reception area? Are there clear signs directing clients to the waiting area? Does the inside feel comfortable? Is the temperature warm enough (or cool enough if it's summer time). Does the paintwork look OK? Are the pictures on the wall suitable? Is the noticeboard up-to-date? Are the leaflets tidy? And go to the toilet. Are they clean? If the inside doesn't seem like somewhere you'd like to go for treatment, leave and visit the next clinic on your list.

So, if you like the location, and you like the look of the place, the next thing to check out is the staff. Do they look professional? How many people work there? Are some more frequent users of rooms? Will you always use the same room or will you have to use whatever is available? Do staff cross refer? Does the acupuncturist recommend the reflexologist etc? Does it feel like a science-based clinic or a 'fluffy' one (ie very little science-based treatments)? Is that what suits you? Can you advertise your services on the noticeboard? Does the clinic have its own Web site? Can you put your information on it? And, how much will it cost you? If you're still giving free initial consultations, you need to take that into consideration when calculating how many hours you want to rent rooms for and how much that will eat into your profits.

Once you have the answers to those questions from all the clinics you have in your area, then you can start to make a choice about which one to choose. Will the clinics that you felt comfortable with let you have the times of day you want? Can you afford to set up in the one you liked best, or do you need to compromise slightly so that you can stay in business? Do you feel the clinic manager was someone you could work with?

Once you've decided on a clinic (or room or whatever) that suits you, your needs, and your pocket, you need to start building the business. That means you need to put on your Web site the details of the new clinic – it's address including postcode, phone number, and even a small map (Google maps makes doing this very easy). You also need to put the days of the week and the times when you will be at that clinic. You want people to easily be able to find you.

It can be very useful to get in touch with other AfSFH hypnotherapists in your new area. That allows you to introduce yourself, and you can learn from them what works in terms of marketing and what doesn't. Sharing experiences and resources can always be valuable. It's also a good idea to have a wander round your new location and get friendly with your local shops, cafes, and hairdressers because they will chat to their clients and can help to spread the word about your new clinic, and they might even allow you to leave your cards in their premises. You might want to offer them free mini-initial consultations and short trance sessions. Hairdressers are generally very chatty and will tell everyone about what they experienced. It's definitely good advertising.

Always have your 'elevator pitch' ready to hand. The more people you meet, the more opportunities you have to briefly describe your hypnotherapy practice. Your pitch needs to be full of enthusiasm, humorous, and factual. You need to appear confident and the kind of person they could trust with their particular issues. You need to make a great first impression – so worth practising.

And you'd be amazed at the power of doughnuts (or even just a packet of biscuits). They are a great way to make new friends – with your new colleagues at your new clinic. And once they've taken one of your doughnuts, you can give them your elevator pitch and before you know it (hopefully) they are recommending their clients to see you.

If business is quiet during those first few weeks, then use the time to write blogs and publish them on your Web site. Google likes Web sites that change regularly and will move you up the rankings. The other thing you can do is make short videos about hypnotherapy. Again put these on your Web site because they help get you up the Google rankings – and that means potential clients are more likely to find you.

Solution Focused Brief Therapy

A look at the history of brief therapy and how it moved the focus from clients' problems to their solutions.

Following the work of people like Freud, a client could be in therapy for a very long time. The thinking was that until you could understand the cause of a problem, there was no way that it could be resolved. And for some people, this more analytical root is still the treatment they seek out.

As the 20th century moved into its second half, people were beginning to wonder whether this approach was the best one to use.

Milton Erickson is one of the originators of brief therapy. Erickson used the analogy of a person who wants to change the course of a river – if he opposes the river by trying to block it, the river will merely go over and around him. But if he accepts the force of the river and diverts it in a new direction, the force of the river will cut a new channel. Erickson also introduced a forerunner to the Miracle Question in which he would ask his client to look into the future and see themselves as they wanted to be, problems solved, and then to explain what had happened to cause this change to come about. A second technique he used was to ask them to think of a date in the future, then work backwards, asking them what had happened at various points on the way.

Similarly, Bill O'Hanlon (who worked closely with Erickson) came up with other ways of getting a client to look to a future without their problem, eg a time machine, crystal ball, rainbow bridge, and a letter from a future self. In one version he would say "let's say that a few weeks or months of time have elapsed, and your problem has been resolved. If you and I were to watch a videotape of your life in the future, what would you be doing on the tape that would show that things were better?" O'Hanlon called his less structured approach Solution-Oriented Therapy and Possibility Therapy.

There was also the Mental Research Institute (MRI) in Palo Alto, California, which used a form of brief therapy that was based on 'the interactional view'. With this approach, problems were thought to happen 'between' rather than 'within' people. Problems would appear when people responded to everyday difficulties in ways that made them worse. The way that a therapist worked was to identify what the 'attempted solutions' were that had caused rather than solved the problems, and then help their clients to do something else instead.

And then, in the late 1970s and early 1980s, at the Brief Family Therapy Center in Milwaukee, Steve de Shazer, Insoo Kim Berg, and their colleagues created the radical new approach of Solution Focused Brief Therapy (SFBT). In addition to the people already mentioned, their ideas built on the work of people such as Gregory Bateson, Don Jackson, Paul Watzlawick, John Weakland, Virginia Satir, Jay Haley and others. Their core idea was that whatever problem a client had come to therapy with, there always seemed to be an exception to the problem, a time when it didn't happen, or happened less or with less intensity. And this led them to believe that the client already had the seeds of a solution and didn't need the therapist to get them to do something different – all they needed was to do more of what they were doing during these exceptional times. The therapist's job was simply to find out what people were doing that was working, then help them to do more of it.

> The OSKAR coaching model uses a solution-focused approach:
>
> Outcome – what the coachee wants to achieve (ie their goal), what it will be like, and what they want to achieve from the session and how they'll know it has been useful to them
>
> Scaling – where the coachee is already, in relation to their goal.
>
> Know how – what skills, knowledge, and attributes (their strengths and resources) the coachee currently possesses that gave them that score.
>
> Affirm and action – affirming is about providing positive reinforcement and reflecting back positive comments about some of the keys strengths and attributes the coachee has revealed. Action is about helping the coachee determine what small action or actions they will now take.
>
> Review – what's the coachee's (positive) progress against actions? What's better? What did they do that made change successful? What do they think will change next? This usually takes place at subsequent sessions.

So, let's look in more detail at SFBT's key assumptions:

- Understanding the cause of the problem is not necessary to resolve it. Attempting to do so may, unwittingly, lengthen or complicate therapy.
- The client's attempted solution (eg avoidance in the case of anxiety) eventually becomes part of the problem. Therefore, changing patterns of response – doing something different – is fundamental to the approach.
- Change happens anyway. However severe the problem, there are times when it is absent, less severe or intense. The therapist must help identify and amplify this change.
- Clients have resources and strengths that can be brought to bear in resolving the complaint. These are often overlooked in problem-focused approaches.
- Clear, salient, and realistic goals are a vital factor in eliciting successful outcomes.
- Poorly-defined or absent goals can prolong or complicate therapy.
- A small change is all that is necessary. Clients are frequently able to manage alone if the therapist can 'start the ball rolling'.
- The client defines the goals and decides when therapy should end.

- Rapid change is possible, even where there is a history of persistent symptoms.
- The relationship between therapist and client is critical; collaboration and a 'robust' working relationship are more important than theory and expertise.
- Each client is unique in their skills, resources, and the way they view their problem. There is therefore no 'one size fits all' solution.
- The focus is on the present and the future, on where the client wants to go rather than where they have come from.
- SFBT sees 'resistance' or hostility as a function of the relationship rather than the permanent disposition of the client.

In the UK, Solution-Focused Therapy was pioneered by Harvey Ratner, Evan George, and Chris Iveson. They established the Brief Therapy Practice, which later became BRIEF. In 2003 this group established the United Kingdom Association for Solution Focused Practice (UKASFP).

Problem-free talk allows clients to talk about what is going well, what areas of their life are problem-free. It can be useful for uncovering hidden resources, and often uncovers client values, beliefs, and strengths. From this, a strength from one part of their life can be transferred-generalized to another area where a new behaviour is required.

> According to Steve de Shazer, although the "causes of problems may be extremely complex, their solutions do not necessarily need to be".

SFBT principally uses questions and compliments to identify a client's goals, and help the client create a detailed description of what life will be like when the goal is accomplished and the problem is either gone or coped with satisfactorily. By identifying 'exceptions', (ie times when some aspect of the client's goal was already happening to some degree), the therapist can help the client come up with appropriate and effective solutions.

SFBT identifies client competencies, ie any behaviours by the client that contribute to moving in the direction of the client's goal. How did they manage to achieve or maintain their current level of progress, are there any recent positive changes, and how did the client develop new and existing strengths, resources, and positive traits?

SFBT uses the acronym MECSTAT, which stands for Miracle questions, Exception questions, Coping questions, Scaling questions, Time-out, Accolades, and Task. While I'm sure we're all familiar with the first four ideas, the last three need some examination.

SFBT uses a time-out to reflect on the developments of the current session. It's preceded by the therapist asking the client if there is anything that the therapist has not asked that the client feels would be important for the therapist to know.

During this break, the client is complimented for their efforts during the session (ie accolades).

The task comes from a brainstorming session where the client suggests behaviours that will help them move towards their goal. The therapist can then ask the client to try this new behaviour – that's their task (what we might call their homework).

As solution-focused hypnotherapists, I think it's very useful to understand the origin and techniques used in Solution-Focused Brief Therapy.

References:

http://www.getselfhelp.co.uk/index.html

http://www.sfbta.org/about_sfbt.html

https://en.wikipedia.org/wiki/Solution_focused_brief_therapy

http://www.solutionfocused.net/what-is-solution-focused-therapy/

http://www.barrywinbolt.com/solution-focused-brief-therapy

http://www.focusonsolutions.co.uk/resources/really_usefull_solution_focused_questions.pdf

http://www.focusonsolutions.co.uk

http://www.uncommon-knowledge.co.uk/training/hypnotherapy/Solution-Focused-Therapy.pdf

Appendix I solution-focused questions

To negotiate goals:

- What needs to happen here today to make you think this meeting was worthwhile?
- What are your best hopes for this meeting today?
- What will be the first signs for you and (others) that things are moving in the right direction?
- What are the benefits for you in achieving your goal?
- What makes you think now is a good time to make some changes?

To identify resources:

- Thinking of a recent difficult time you've had, how did you manage to come through it?
- What did you/ they find helpful?

- What was unhelpful?
- How do you/ they talk themselves through difficult situations?
- What do you know about yourself (or your situation) that reassures you that you can deal with this?

To motivate:

- On a scale of zero to ten, with ten meaning that you would do almost anything to achieve your goal and zero being hardly anything, where would you put yourself today?
- How confident are you that you already have the ability to achieve your goal?
- Can you find evidence for this?
- How will you know that this is a good time to make a start at least?

To focus on the future:

- Imagine one night when you are asleep, something amazing happens and the problems that have been worrying you disappear. Since you are asleep you don't know that this has happened. When you wake up in the morning and go about your day, what will be the first signs for you that things have improved?
- Do you think that any of these things have happened to you recently, even for a short while?
- If I were to meet you in a few months' time and you were to tell me that things were getting better, what would you tell me had happened?

To maintain progress:

- What do you think you need to do/think/remember to give you a chance of keeping on your solution track?
- What will you say to yourself if you have a set back?
- What will be the best way of handling a set back?
- What will be helpful / unhelpful?
- How will other people know that you need help at that point?
- How will you overcome any obstacles that come in the way?
- Should plan A not work, what's plan B?

To increase self-awareness:

- What does it say about you that you managed to do that (when person has achieved something)?
- Did you know you could do that?

- Where did you learn to do that? Have you done it before?
- Has anyone else noticed that you have done this?
- What do they think about it?

To develop strategies:

- What could you do about the things over which you do have control and how could you accept the things you can't change?
- What have you tried so far that you know does not work?
- Do you know anything that worked for anyone else in this situation?
- How would you advise a friend who was having this problem?
- What is the smallest step you could take in the next forty eight hours that would be helpful?

Making changes

A look at different models of how the brain, body, and environment interact and how to help a person make positive changes to their life.

Before you can make changes in yourself or a client, it's useful to understand how you got to be the way you are now, and also get an idea of what techniques for making change are available and which work more successfully than others.

It's easy to think of what you know as 'you' as simply residing in your head. It's quite easy to think that this is where it all happens. This is where you think of things, it's where you feel emotions, and it's where your memories and your belief systems are located.

But you are also hugely affected by your body. To start with, there are your genes, there are the hormones and how much of each is present in your blood at any one time. There's what you eat and it's effect on your gut biome. There's how much exercise you do and how much sleep you get. There are things like how effective your immune system is. And much more.

There are also external factors like how much stress you are under. There are social and environmental factors that affect you. There's your family background, your education, your interactions with your family and work. There's the social condition you find yourself in at the moment, and there's much more.

Let's take a look at these factors in more detail. Let's start with the brain – that three pounds of neurons and glial cells. The brain receives messages from our sensory organs, which includes sight, hearing, taste, smell, and touch, plus temperature, movement, knowing the relative positions of the parts of the body, pain, balance, vibration, and there's a whole lot more like hunger, needing to go to the toilet, etc. The brain even has a sense of time passing. These feed information into the rest of the brain, which uses the information to build a model of the world that it can use to predict what will happen next in various situations. It's important to recognize that this model is only an approximation to the real world and can be updated as new information is received (although some people prefer to ignore any new and contradictory information).

The brain can make the body move using the somatic nervous system and the autonomic nervous system. The somatic nervous system is under voluntary control, and connects the brain to muscles. The autonomic nervous system influences the function of organs over which we have little or no voluntary control. The autonomic nervous system is divided into the sympathetic system and the parasympathetic nervous system. The sympathetic nervous system is used in 'fight or flight' situations. The parasympathetic nervous system allows the body to function in 'rest and digest' mode.

Robert Dilts (based on some work Gregory Bateson) developed an idea of neurological or logical levels of how the mind works (see Figure 1). He envisaged six different levels. The top level is the slightly strange 'spirituality/connectedness', which represents a person's higher purpose/contribution to the world. The next level is

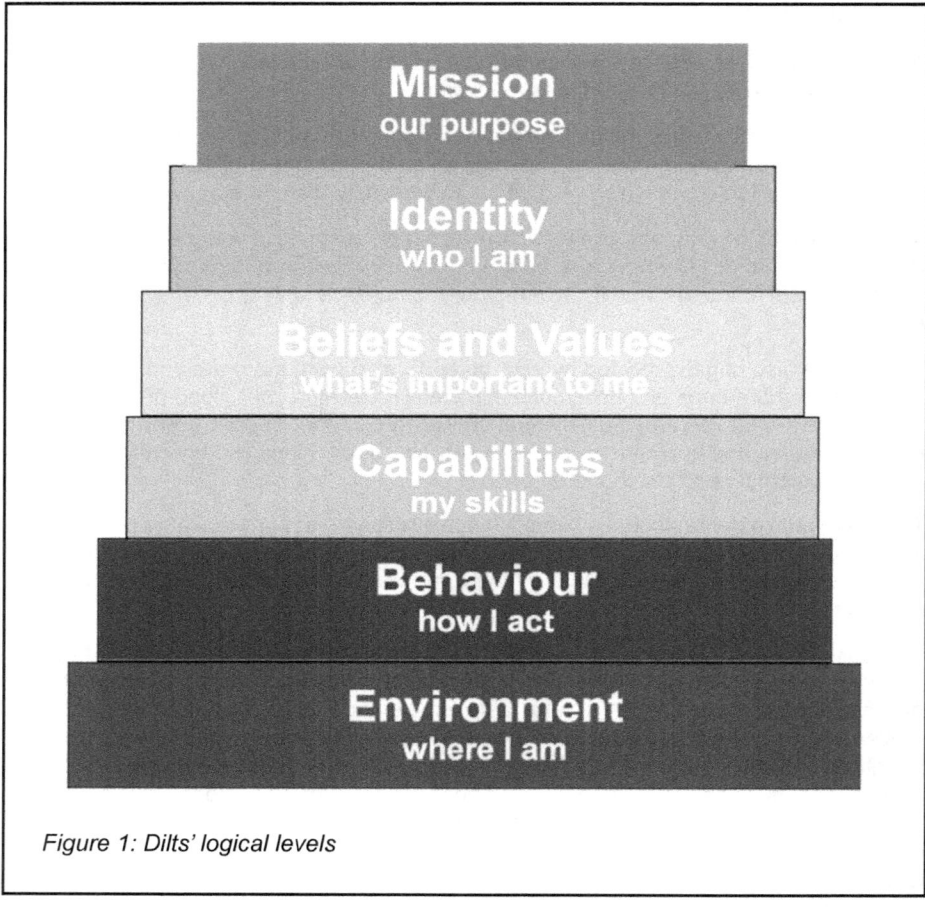

Figure 1: Dilts' logical levels

'identity'. It's all about a person's sense of who they are and their roles in life. The level below that is 'beliefs and values', and they influence the way we think and act.

Values are how we decide whether our actions are good or bad or right or wrong. Values are arranged in a hierarchy, with the most important one at the top and the others below that. Values are those ideas in which we are willing to invest time, energy, and resources to either achieve or avoid. A person's values can change with different contexts and when they are in different environments. Beliefs are presuppositions that certain things are true or real. A limiting belief is a belief a person has that they can't do something – like speak in front of an audience. Beliefs can change if a person's skill set changes. Beliefs are created depending on what events have happened to us in the past. Values and beliefs usually operate without people being aware of them, but they have a big influence on how people behave. Attitudes are collections of values and belief around a certain subject. People are typically aware of their attitudes, and can identify how they feel about different topics

The next level is 'capability', which is to do with the skills, abilities, strategies, talents, and resources that guide a person's behaviour and enables them to take action. It's about how we do things and the skills and processes that let us know we can carry out a task or act in a certain way. When we start to learn a new skill, such as playing the banjo, this layer is empty, but it gradually fills up until we can play the instrument (or whatever) without having to think about it.

'Behaviour' is the next level, and it refers to how people act in a particular environment. These actions include: thinking, speaking, listening, reacting, and consciously acting in a particular way in order to achieve a goal. It also includes what people don't do – eg not responding in a particular situation.

The last level is 'environment' and is to do with your physical environment, who's with you, your cultural group and the constraints that group puts on you. This is clearly part of our third group – external factors.

With NLP, a change that happens at a higher level usually has some impact on the lower levels. However, sometimes a change doesn't take place, and it's useful to look at each of the neurological levels to see where exactly the block to the change occurs.

We usually focus only on changing behaviour, but with these logical levels, we can see there are other questions to ask. In terms of 'environment', the question is "where do I need to change?". With 'behaviour', the question is "what do I need to change?". At the 'capability' level, it's "how do I make these changes?". For the beliefs and values level, the question is "why do I make these changes?". And at the 'identity' level, it's "who am I and do I reflect that in the way I live?".

When thinking about the brain, people often refer to emotions, moods, and a person's temperament. Moods are generally less specific and intense than emotions, and people generally describe themselves as being in a good mood or a bad mood. Moods differ from emotions, feelings, or affects in that they are less specific, less intense, and less likely to be triggered by a particular stimulus or event. Having a client in a good mood is probably better when talking about change.

Temperament refers to aspects of an individual's personality (eg introversion or extroversion) that are often regarded as innate rather than learned. There are numerous classification schemes for temperaments, including the four humours (sanguine, choleric, melancholic, and phlegmatic) and Thomas and Chess's nine temperament traits in children (activity, regularity, initial reaction, adaptability, intensity, mood, distractibility, persistence and attention span, and sensitivity). As you can see, temperament is kind of vague and therefore not useful when helping clients make changes.

Willpower is usually associated with the brain. It's the capacity to restrain our impulses, to resist temptation, and to do what's right and good for us in the long run, not what we want to do right now. It's what we use to control and manage our thoughts, impulses, and emotions, and it helps us persevere with difficult tasks. A person can build up their willpower by exercising self-control regularly in small ways. But willpower can get tired if it's used too much. If you have spent a lot of time using your willpower, you will not have much left if another challenge comes along soon after. Exercise, a good night's sleep, and eating properly can all help to build up a person's level of willpower. It's worth making this clear to a client when discussing making changes in their life. If they are feeling that stopping smoking is hard and it's taking all their willpower, there will come a time when they will 'run out' of the strength to keep on. That's why, the client has to be totally committed to making a change in their life, otherwise it will be very hard for them to sustain, despite our best efforts.

Things like mindfulness, meditation, prayer, and chanting or singing, affect the brain and can have a big impact on how people view themselves and events going on around them. The filter that we apply to messages arriving at the brain affects how we respond to the event – from being able to shrug it off to feeling highly emotional about it and having our behaviour driven by that emotional response.

Our bodies are affected by what we eat and drink in terms of diet and medication, how much exercise they get, how well we sleep, stimulation of the vagus nerve, the type and health of our gut bacteria, our genes, what hormones are flowing through our body, how well our immune system is working, and other internal factors. How we stand, move, and act, can also impact on how we feel and how we behave. Doing things like tai chi and yoga can impact on how well our bodies feel. And, obviously, illness affects our body and what we think we can do.

All sensory nerves (apart from smell) terminate in the thalamus. Various estimates suggest that up to 13 million messages arrive at the thalamus each second. Clearly, there's no way that the brain can deal with all of those. Before messages are sent on to the rest of the brain, they are filtered – in effect, they can be deleted, distorted, or generalized.

Deletion allows us to ignore some of the messages. It brings the number down to a more reasonable amount. Distortion occurs when we misrepresent reality by changing the sensory data. This makes it easier to accept and requires the brain to do less thinking. Generalization is like inductive reasoning. We take a couple of examples and draw a general rule about the world. Again, it's a way of saving on brain activity.

We know there is feedback from the body to the brain. So, if we are standing slumped against the wall, for example, our brain will think we are feeling tired and perhaps depressed. Similarly, if we are standing upright and looking up at the world, our brain will interpret that as feeling confident and being full of energy. One of the techniques we use to make changes in clients is to get them to act 'as if'. So if they want to feel more confident giving a presentation, we get them to appear to be confident. And the feedback loop will tell the brain that they are confident.

Colours and moods:

- Red – passion
- Orange – welcoming, energy
- Yellow – happiness, positivity
- Green – harmony, stability
- Blue – peace, relaxation
- Purple – luxury, romance
- Black – power, elegance
- White – purity, simplicity
- Brown – dependable, friendly

External factors include: how bright the lighting is; what colours you see around you and what images you look at; what ambient sounds you can hear or what music you're listening to; whether you can see the countryside (health); hearing rousing speeches; group agreement; peer pressure; seeing others (Bandura/mirror neurons) do things; how much touching or cuddling you experience (affects oxytocin levels); having a massage; listening to or reading logical argument; hearing emotional argument; hypnotherapy; writing down a goal; continually visualizing goals and the steps needed to achieve those goals; how much stress you are under; your family background, your education, your interactions with your family and work; and much more.

We know that if a client wants to be more active or lose weight, it's best if they spend their time with people who go to the gym or out running and join in with those activities and conversations. The clients want to stop spending time with people who tend to sit down all the time and eat chocolate bars throughout the day.

> A study, published in 1984 in the journal *Science* by Roger Ulrich, was the first to demonstrate that gazing at a garden can sometimes speed healing from surgery, infections, and other ailments.

> Albert Bandura's Social Learning Theory suggests that people learn from one another, through observation, imitation, and modelling.

Similarly with clients who are stopping smoking: they need to avoid spending time with smokers and spend time with people who don't smoke. That will help them break the smoking habit.

Changing environment had a big impact on US soldiers during the Vietnam war. Many soldiers were taking drugs, but most came home and stopped the habit immediately. You also find people in hospital, who take morphine to control the pain, can come out of hospital without having a drug habit.

> A mirror neuron (assuming they exist) is a neuron that fires when an animal acts and also when an animal observes that same action being performed by another.

Clearly, these three elements (mind, body, and environment) interact.

Cognitive Behavioural Therapy (CBT) has its own model of how these interact (see Figure 2). CBT has two key principles. They are:

- Our thoughts influence the way we feel and behave in any given situation

- Our interpretation of a situation is influenced by the beliefs we hold about ourselves, other people, and the world around us.

So, when we feel depressed or anxious, our interpretation of a situation becomes negative or fearful, which prompts us to avoid, withdraw, or escape from the situation. These behaviours simply maintain the problem. By changing the way we interpret a situation or event, we change our thoughts, which, in turn, change our feelings, behaviour, and the physical reaction or response of our body.

As mentioned earlier, any event that takes place around us is filtered (through, what's called, a schema) and interpreted before it reaches our thoughts. A schema is like our model of the world, so if an event fits our model, we let the information through and

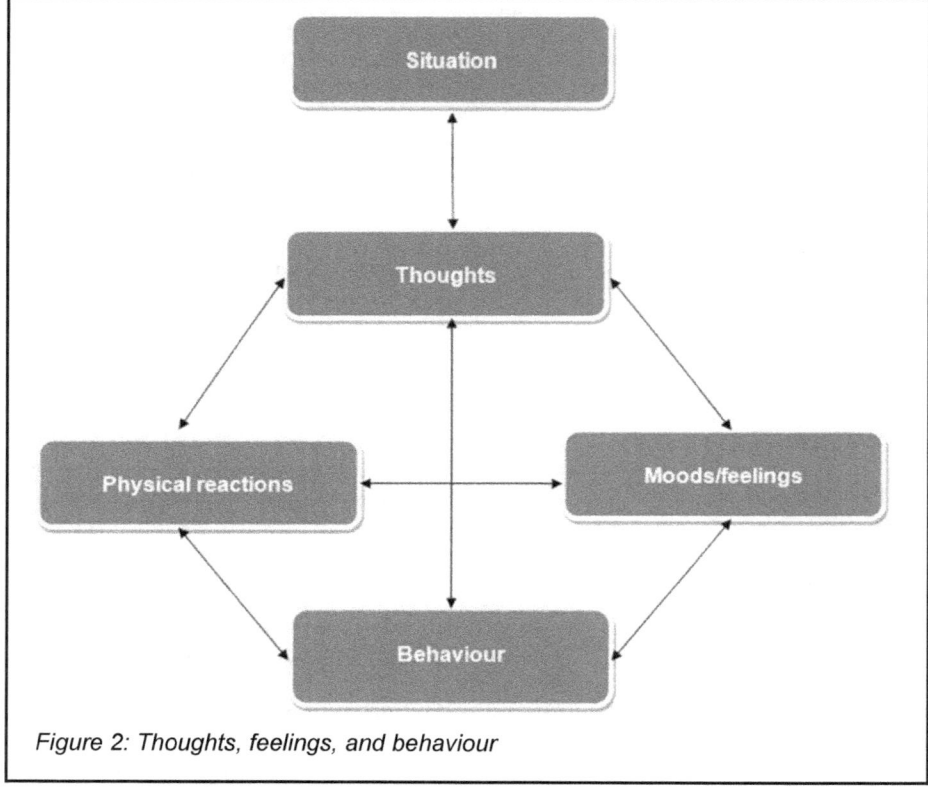

Figure 2: Thoughts, feelings, and behaviour

deal with it. If an event doesn't fit our model, we tend to distort the information (to make it fit) or delete it (ignore the event completely).

CBT also uses the idea of 'automatic thoughts'. These are thoughts that occur whenever a particular situation is experienced. Individuals may or may not be aware of their automatic thoughts, but people can usually learn to be aware of them. The thoughts can be negative or positive, but we usually see clients with automatic negative thoughts. It's possible to help people make changes by helping them to become aware of their automatic negative thoughts. Firstly, they should write down the emotions they were feeling when a particular incident (activating event) took place. They also need to rate the intensity of those feelings. Secondly, they need to write down the actions that they took. Thirdly, they should write down what triggered their thoughts and feelings (the activating event, eg people objects, images, etc), Next they must write down their negative automatic thoughts and their beliefs about the event – including how it seemed to them and what it meant to them.

The next stage is to look at the negative thoughts and beliefs and see whether alternative explanations are possible. They should label the thinking error for each distorted thought and belief. Then for each of the original thoughts and beliefs they wrote down, they should create a balanced and realistic response and write that down. Lastly, they need to re-rate their original emotions and jot them down. Then, write down a plan for more constructive behaviour, eg a different way of managing the situation in the future.

Our emotional intelligence has a big impact on how we behave with other people. Emotional Intelligence (EI) is the capacity of individuals to recognize their own, and other people's emotions, to discriminate between different feelings and label them appropriately, and to use emotional information to guide thinking and behaviour. That means if you can help a client increase their emotional intelligence, it's possible to make changes in their behaviour and also their thinking.

You can think of these models and techniques as being tools in your tool bag that can be called upon when needed to help a client make changes, but they're not as powerful techniques as hypnotherapy is. Hypnotherapy might be thought of (metaphorically) as a hammer and some screwdrivers. And solution-focused hypnotherapy gives you a power drill to use. A lot of the other techniques can be thought of as children's versions of these tools – they can be useful in a way, but not as good as the real thing. But there comes a time when a pair of pliers or some spanners are just what you need. And that's when some of these additional techniques really come in handy.

A lot of the work we do to help people make changes is simply to overwrite their old 'bad' habit with a new positive one – one that fits with their desired goal (like stopping the smoking habit). In his book, *The Power of Habit*, Charles Duhigg suggests that a habit is actually a three-part system, what he calls a habit loop (see Figure 3). There's a trigger (the situation that sets it off); the behaviour; and a reward (why your brain wants you to do the same thing again next time). Helping your client to clearly identify the trigger for their old habit will help them to stop repeating it (if that's what they want to happen).

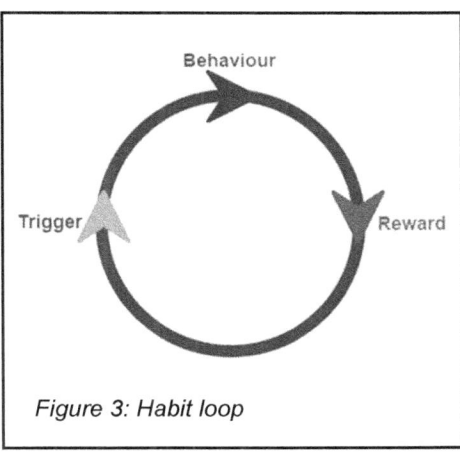

Figure 3: Habit loop

B J Fogg, Director of the Persuasive Technology Lab at Stanford University, suggests that new habits are hard to create because we tend to sabotage our own efforts. He suggests that we should define a first step (going to the gym, etc) that takes less than sixty seconds to do. So, rather than trying to build a habit of going to the gym, your client builds a habit of picking up their kit bag and getting in the car to drive to the gym as soon as they get up. It's a micro-habit, and big habits are made up of these micro-habits. And as the client successfully completes each micro-habit, they've established their desired larger habit.

This leads our clients to an easier way of creating new habits, provided they can identify the triggers or cues that cause the habit. They need to know exactly what the trigger is – what it looks like and when it occurs (a bit like hot thoughts in CBT). Usually, this leads them to a particular behaviour – the one that you are trying to stop or modify. They need to be able to describe the behaviour, and, if possible, say what benefit they get from it. The last stage is to say what they will do instead. Now remember, this can't be some vague aspirational habit like 'go to the gym', it needs to be a clear and definite micro-habit – one that is probably the first step in a larger habit. So, you want your clients to be saying, "When x happens, instead of doing y, I will do my new micro-habit".

Sometimes, you may need to persuade a client that now is the time to make a change that they have been talking about. Researchers led by Maurits Kaptein found that people are most likely to be influenced when they receive the right message, at the right time, and in the right way. Let's unpick that. The right message is one that they are receptive to because they generally agree that the goal is worthwhile. This is easy, because they have come to see you to stop smoking, they are going to be receptive to that message. People need to hear the message at a time when they're likely to be influenced by it. Again that works for us, but does explain why there's no point trying work on other parts of a client's life until they are ready. So, no point talking about stopping smoking until they have made that decision because they won't be receptive. Thirdly, the message itself has to present a person with data that they will believe. Again, the initial consultation and refreshers give them plenty of facts that they can check with a Google search and so will believe.

Kaptein and his team were able to identify seven factors that influence people to make changes in their life. They were:

- Authority – people are inclined to follow the advice of those they view as authorities.

- Consensus – if people see others behaving in a particular way, or adhering to a certain belief, they are more likely to do the same. This is how Weight Watchers works, and why giving clients examples of similar clients at similar stages successfully making changes works.

- Consistency and commitment – people like to appear consistent and to behave in ways that reflect their underlying beliefs in what's right or good. So, a client who believes that hypnotherapy helped them for one thing are likely to book to see their therapist for another problem.

- Scarcity – when something seems to be unavailable, people want it more. That's why being booked up all next week, makes clients want to see you more than being available later today. It might also explain why higher prices work – you can only see a few clients who can afford that price.

- Liking – if a person likes someone, they are more likely to agree with the message that person communicates. That's why it's so important to build rapport with a client before starting work with them.

- Reciprocity – if someone does something nice for another person, they are more likely to want to respond in kind. This works in marketing when a person is given a free sample, they are more likely to buy. It may explain why free initial consultations work (although I'm not sure on that one).

The third finding from the team was about personality characteristics that affect how susceptible a person is to being persuaded. They categorized people as:

- Having a high need for cognition – these people scrutinize all incoming messages and can be sceptical of so-called 'authorities'.
- Being conscientiousness – these people are likely to go along with messages that rely on consensus or social proof.
- Having a desire for consistency – these people, having agreed over a small request, will be more likely to agree to a larger one – so they appear to be consistent.

You'll have to decide on the personality of your client to see which one of these will work the best in terms of them making the changes they want.

One great trick of persuading someone is priming. This is here you either hide the words you want them to hear in other words (eg by saying 'stop and 'smoking in separate sentences), or you use words close to the desired word (like saying doctor when you want them to say nurse). Basically the person is halfway to your desired response.

Yale University found that as well as being credible, a speaker needed to be attractive in order to persuade people. And the most persuasive words to use, when persuading someone, are: 'you', 'because', 'free', 'instantly', and 'new'. Some useful phrases to use when influencing and persuading clients are:

- "What if" – this phrase removes the ego from the discussion and creates a safe environment for curiosity and brainstorming.
- "I need your help" – this changes the roles of dominant and subordinate, engaging the other person and providing a transfer of power.
- "Would it be helpful if" – this phrase changes the focus from the problem to the solution.

Some other tips for getting clients to agree with you are:

- Avoid arguments.
- Show respect for the other person's opinions. Don't say, "You're wrong"
- If you are wrong, admit it quickly and emphatically.
- Begin in a friendly way.
- Get the other person saying "yes, yes" immediately.
- Let the other person do most of the talking.
- Let the other person feel that the idea is theirs.
- Try honestly to see things from the other person's point of view.
- Be sympathetic with the other person's ideas and desires.

- Appeal to the nobler motives.
- Dramatize your ideas.
- Talk about benefits instead of features.

And, of course, sometimes you have to write persuasively. Many hypnotherapists write blogs that act as long adverts for the services they offer. Here are some tips for how to do that successfully:

- Be definite
- Be positive
- Repeat yourself
- Be personal (use words such as 'we' or 'I')
- Use questions (they make people think)
- Use feelings (emotionally-charged words).

You're probably familiar with used car salesmen using many of these techniques. But they can be used less overtly and more subtly – and, in fact they are. You just have to watch a group of people making a decision (like where to go for dinner) to see some people using these techniques to successfully persuade the others in the group. And you can see people doing the opposite of these techniques failing to persuade anyone.

So, armed with this information, you can use these techniques to help convince your client that the changes they feel they ought to make at some time in the future, should be made now.

References:

http://www.fastcompany.com/3030173/work-smart/how-to-use-10-psychological-theories-to-persuade-people

http://www.bbc.co.uk/bitesize/ks3/english/writing/argue_persuade_advise/revision/3/

http://blog.enhancv.com/8-persuasion-techniques-to-change-anyones-mind/

https://www.psychologytoday.com/blog/fulfillment-any-age/201506/7-ways-people-can-change-your-mind

Hypnotherapy and exam nerves

What's going on inside us when we're feeling nervous about an exam (or any other event)? How can hypnotherapy help reduce the nervousness and help the candidate to perform to their best ability?

The revision calendar has been drawn and colour-coded, the actual revision has started, and now the first exams are looming up on the horizon – and your poor exam candidate is finding themselves turning to jelly. They can't learn anything and they can't remember what they've already learned. And they're running out of time. And these exams are the big ones – they need to do well. Other symptoms that students experience include worry, irritability, tears, sleepless nights, loss of appetite, depression, headaches, and stomach aches. Can hypnotherapy help?

Before we look at the ways that hypnotherapy can help, let's look quickly at what's going on inside the examinee's brain.

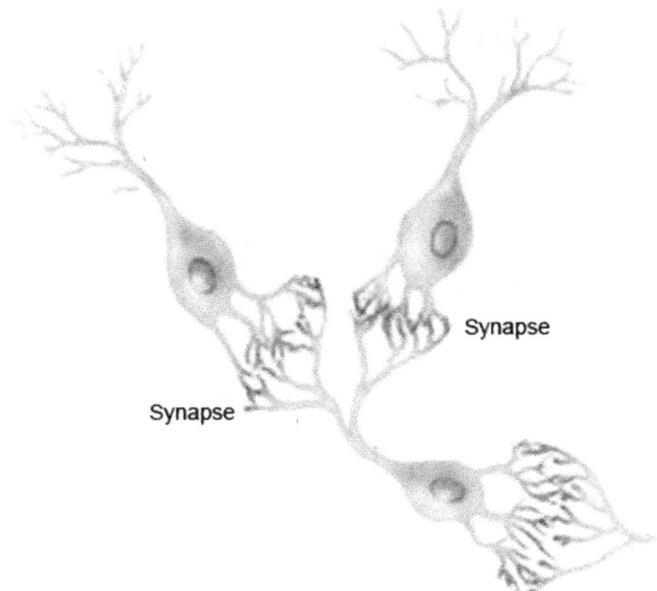

Figure 1: Connecting neurons

Your brain is about three pounds of jelly (glial cells) with nerve cells (neurons) inside it. The number of neurons you have isn't fixed, you can make more. And the neurons are connected, although not directly. There are gaps (synapses – see Figure 1) between the neurons. The more connections there are between neurons, the stronger a link is made. This explains why we always carry out our 'habits' because the links in our brain between the neurons are so strong. New learning, whether that's a memory of an event or the enforced learning of the history syllabus, involves new connections being made or existing ones being strengthened, ie more-and-more little dendrites from one nerve cell connecting to the dendrites of a second nerve cell. It may also involve new neurons being used. This is part of the natural plasticity (ability to change) of the brain.

Exercise is very good for creating Brain Derived Neurotrophic Factor (BDNF) and this helps with the growth and differentiation of new neurons and synapses. You can find it in the hippocampus, cortex, and basal forebrain, which are all areas vital to learning, memory, and higher thinking.

The brain can be divided up into different regions (see Figure 2), like the lobes; but if you look internally, you can see that there are three clearly defined regions (see Figure 3).

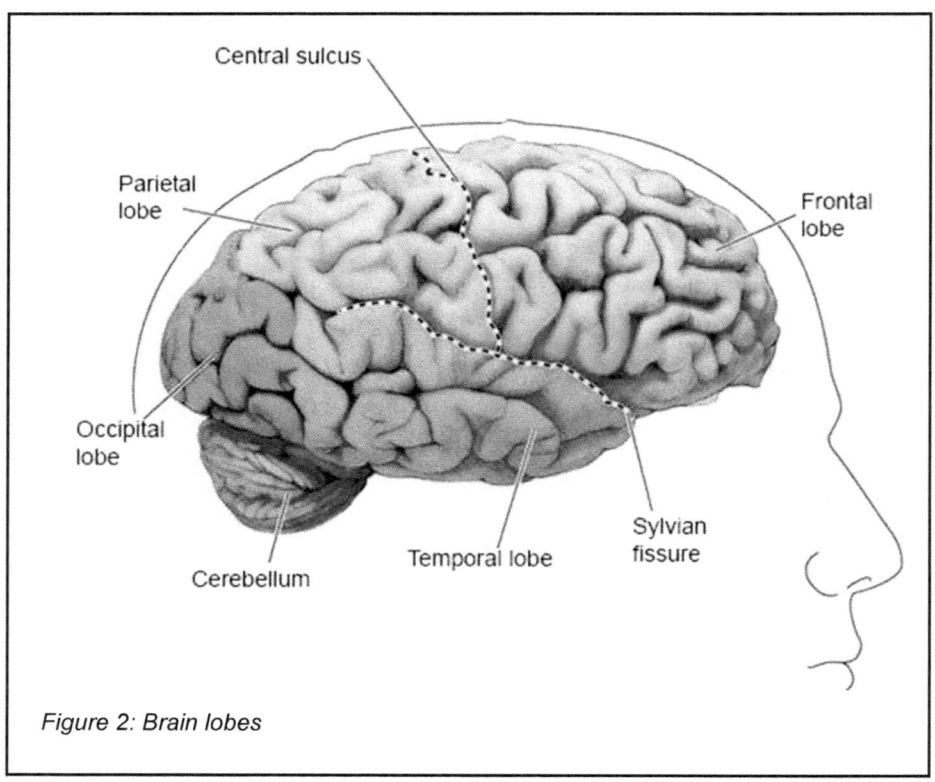

Figure 2: Brain lobes

There's the brainstem, which is responsible for telling the heart to beat, the lungs to breath and the GI tract to squeeze food along it (amongst other things like controlling the iris in the eye). The limbic system looks after the 4Fs (fighting, fleeing, feeding, and reproductive behaviour). And higher functions (such as problem solving, maintaining attention, and controlling emotional impulses from the limbic system) are handled by the cerebral cortex. The limbic system is very fast and the cerebral cortex is much slower and tends to be used less.

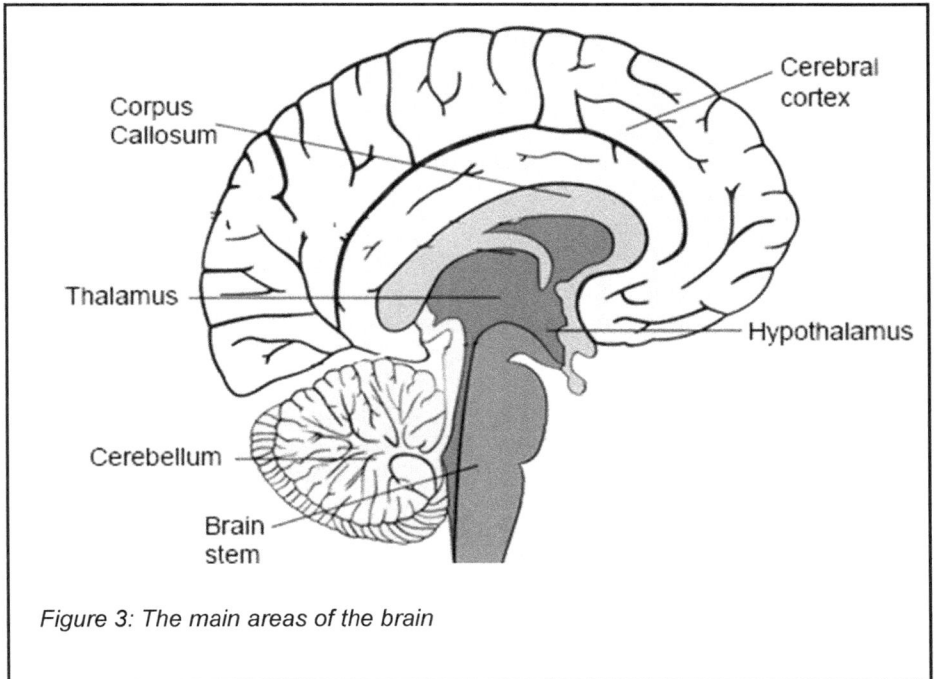
Figure 3: The main areas of the brain

In the limbic system, you'll find the thalamus, hypothalamus, amygdala, and hippocampus. The amygdala is associated with fear. And when you're fearful you can't access other parts of your brain. It's called amygdala hijack. You may remember times when your maths teacher stood threateningly over you and you couldn't remember even the simplest things. Hypnotherapy can help with ways to become calm and stay calm.

The way that amygdala hijack works is this. Messages from the body go to the thalamus. From here, part of the message goes to the amygdala and part is sent to the cerebral cortex. If the amygdala perceives a match to

> There can be positive hijacks too, eg when a joke seems so funny that a person's laughter is almost explosive.

the stimulus, ie the hippocampus contains a memory of this type of stimulus being a threat, the amygdala triggers the HPA (Hypothalmic-Pituitary-Adrenal) axis and hijacks the rational brain. If the amygdala doesn't find a match, the body follows the logical instructions from the cerebral cortex.

Going back to the HPA axis: the hypothalamus produces Corticotropin-Releasing Hormone (CRH). This (with vasopressin) stimulates the pituitary gland to produce ACTH (AdrenoCorticoTropic Hormone). This travels in the blood to the adrenal glands where it triggers the production and release of corticosteroids and cortisol from the (cortex of the) adrenal glands. These chemicals put the body into a high 'readiness' state. Heart rate, blood pressure, and respiration rise so that muscles and the brain are supplied with more blood and, consequently, more oxygen. Blood flow decreases to the stomach, kidneys, skin, and liver. Sexual and immune functions are suppressed. Endorphins (natural opiates) are released to relieve potential pain. And fats and sugars are released into the blood stream to supply extra energy.

The sympathetic nervous system is stimulated. Sympathetic fibres that end in the adrenal medulla secrete acetylcholine, which activates the secretion of adrenalin and to a lesser extent noradrenalin from the adrenal medulla. They increase the heart rate and respiration rate, and raise blood pressure. Adrenalin and noradrenalin positively feed back to the pituitary gland.

Fight and flight burns up the glucocorticoids – and once the person is safe, the body's chemistry returns to normal. The immune system starts to work again as the body rest and repairs itself. Note that glucocorticoids act on the hypothalamus and pituitary to suppress CRH and ACTH production. This is a negative feedback cycle. If we're stressed and we don't fight or flee, the chemicals stay in our body and can do damage. Exercise is like fighting or fleeing and so will remove these hormones from the body – which is good.

The hypothalamus also produces Thyrotropin-Releasing Hormone (TRH), which stimulate the thyroid gland to secrete thyroxine. Thyroxine controls the rate of metabolic processes in the body.

The more a student worries about exams, the more they fill their metaphorical stress bucket. As that bucket gets fuller, it becomes harder to cope with life's stresses. And, if you've ever had a panic attack or any other kind of 'melt down', that's a consequence of your 'stress bucket' overflowing.

Hypnotherapy can help you to sleep better, which is when the body naturally empties its stress bucket. That's why working all through the night is counterproductive when it comes to learning new material. If you're studying hard, those memories are consolidated during sleep and that helps you to be able to recall material in the future. If you lose sleep, you are not as 'clever' the next day (cognitively disadvantaged).

The HPA can also kick in on exam days. Hypnotherapy can help students with techniques that can help them stay calm. Of course, a certain amount of nervous tension is not bad. A small amount of adrenalin can help exam performance by keeping students alert and focused. It can even help them to think faster.

Eating is also important. Make sure that learners have enough food, but not too much, otherwise the body wants to go into rest and digest mode – which isn't the ideal for answering exam questions Students also need to make sure that they drink enough water. Being dehydrated is not a good way to learn.

Because young people at school are being constantly tested these days, there's another unfortunate consequence. When we are in a fearful situation (the last lot of exams), we retain the memory of how bad it was and what happened to us (a bit like becoming phobic about going on a plane). So, the next time we start revising for an exam, we remember what happened the last time and we tend to repeat our behaviour

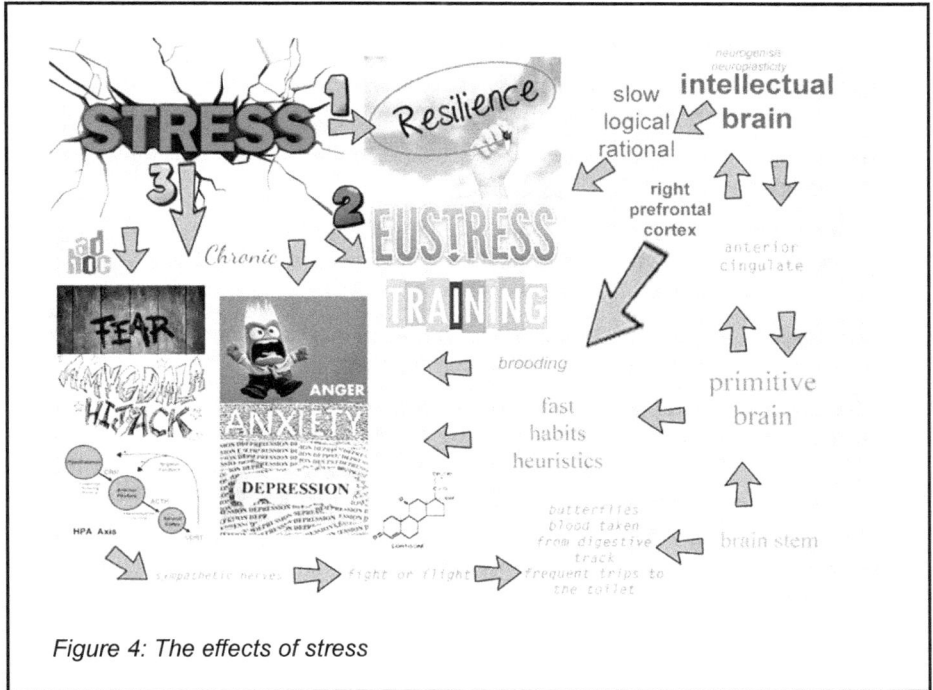

Figure 4: The effects of stress

(panicking, headaches, not being able to sleep), which makes it harder to do well in the current exams. As solution-focused hypnotherapists, we can use the rewind and reframe techniques to help remove these emotions and behaviours.

Figure 4 summarizes all this. Some stress (1) you can deal with using your natural resilience. Some more stress (2) makes you stronger (like when you're training for an event). But even more stress (3) is bad for you. Short periods of this extra (*ad hoc*) stress often results in a fear response including amygdala hijack and the HPA (hypothalamus, pituitary, adrenals) axis being invoked. This is your fight or flight response, the one where you lose blood from internal organs so that your muscles have plenty of energy – which, consequently, leaves you visiting the loo a lot! If the extra stress continues for a longer time (chronic), you probably start to experience anger, anxiety, and/or depression, and your body has lots of excess cortisone.

Your intellectual brain can decide how stressful an event is – but while being logical and rational, it is also slow. So the faster primitive brain can use its heuristics (rules of thumb) to decide which habit to invoke (eg running away screaming). You can get used to events (acclimatize) and your anterior cingulate can keep a lid on the primitive brain's response.

If you brood on events (right prefrontal cortex), you effectively create chronic stress – and get the usual result.

Sympathetic nerves help create the fight and flight response. The more relaxed 'rest and digest' response is controlled by the parasympathetic nerves.

There are a number of techniques that a hypnotherapist can use to help a student (and other people) deal with stress. These include:

- Bucket emptying – helping the student to remove the stress that they have added to their metaphorical stress bucket.
- Confidence building/relaxation – helping the student to feel more confident about their abilities and more relaxed about the exams they are facing. Getting them back in the intellectual brain and more in control of the decisions they make.
- View things differently – looking at what's ahead of them in a more realistic way so they feel less stressed by the demands being placed on them.
- Reframing and future pacing – getting the student to visualize themselves successfully revising and getting lots of work done. Get them to see themselves successfully remembering the information they need to know. This works better than simply picturing themselves holding up a certificate with lots of 'A's on it.
- Rewind – removing the fear from any memories they may have of not being able to revise or not having successfully revised in the past.

The following techniques can be used on exam days to help calm down the student if they find they are beginning to feel nervous:

- Anchoring – creating a link between a simple action (like rubbing their ear) and the feeling of being relaxed and in control.
- Peripheral vision relaxation – using the calming effect of the parasympathetic nervous system to help relax.
- Circle of excellence – creating an imaginary area that has all the skills needed to be successful at a task. The student simply imagines stepping into the circle to feel fully equipped for the task ahead of them.
- 7-11 breathing – breathing in for the count of 7 and breathing out for the count of 11 also uses the parasympathetic nervous system to help calm down.
- Mindfulness urge surfing – a technique of recognizing that you feel a certain way at the moment, and this feeling may get worse for a short period of time, but after that, it will get better.

It's perfectly normal to feel a degree of fearfulness when facing a challenge. Using solution-focused hypnotherapy can not only help with the stresses faced on exam days, it can also help a student to cope with the stresses of revision, and this can help with improving memory and recall, and the way that exam questions are answered.

References:

https://en.wikipedia.org/wiki/Hypothalamic%E2%80%93pituitary%E2%80%93adrenal_axis

https://en.wikipedia.org/wiki/Synapse

https://en.wikipedia.org/wiki/Limbic_system

https://en.wikipedia.org/wiki/Amygdala

http://www.exeterexpressandecho.co.uk/Hypnotherapy-help-tackle-exam-nerves/story-29080791-detail/story.html

Dealing with panic attacks

Here are some ways to help clients who are experiencing panic attacks.

It's not uncommon to have people who have experienced panic attacks, or who expect to have a panic attack, come to see us. Panic attacks seem to be on the increase, and hypnotherapy – with our usual techniques of bucket emptying and increasing a client's confidence – can be a great way of helping people get over their panic attacks. But what actually is a panic attack and what other techniques can we teach our clients to help them to stay calm in situations that previously would have caused a panic attack.

Usually, a panic attack lasts between 5 and 20 minutes. It occurs when your body goes into fight or flight mode. Your body needs more oxygen, so your respiration rate increases. The release of adrenalin into the blood increases your heart rate, takes blood away from your intestines, and causes your muscles to tense up. Other symptoms that people report include:

- Not only a faster heartbeat, but also an irregular heartbeat (palpitations).
- A shortness of breath (hyperventilating)
- Sweating
- Trembling/feeling shakey
- Nausea
- A choking sensation
- Dizziness
- Tingling fingers
- Ringing in your ears.

Some people report that it feels like they are going to have a heart attack or that they are going to die. Other people report feeling disoriented or having a dry mouth.

And other people report panic attacks taking completely different forms. They suggest that, for them, a panic attack looks like:

- Unpredictable bouts of rage or irritability
- Nit-pickiness (obsessive behaviour, which may be a part of OCD), and even a hypersensitivity to disarray, chaos, or any sort of change
- Fast-talking, stuttering, stumbling over words
- Not talking at all
- Sitting rigid, staring into space, almost seeming 'zoned out'.

Panic attacks often occur in a trigger situation, such as when you are arriving at an airport for that much dreaded flight. But often there isn't a trigger, they just seem to occur for no obvious reason. And there are reports that they can happen while people are asleep – these are called nocturnal panic attacks.

So, as I said, apart from bucket emptying and confidence building, what can you do for someone who experiences panic attacks? The answer seems to be breathing exercises, grounding, and perhaps mindfulness.

Two breathing techniques that work well are 7-11 breathing and square breathing.

7-11 breathing involves breathing in for a count of 7, then breathing out for a count of 11. When breathing in, use 'diaphragmatic breathing' rather than shallower higher lung breathing. And if the timings seem too long, count faster! Most people feel calm after 5 minutes or so of 7-11 breathing.

With square breathing, a person can sit upright in their chair with both feet on the floor. And then the procedure is:

- Inhale 2 3 4
- Hold 2 3 4
- Exhale 2 3 4
- Hold 2 3 4

The person needs to focus on their breath and the count of four, and keep repeating the breathing pattern until they feel calm and relaxed.

Grounding exercises provide a way to anchor a person in the present. The idea behind it is to use a person's senses (sight, hearing, smell, taste, and touch) to link the mind and body in the present moment and turn off the growing panic response. Some techniques that have been suggested for grounding include:

- Talk about yourself. Say your name, your current age, where you are now, what you have done so far today, and what you plan to do next.
- Stop, look, and listen. Notice and name all the sounds you can hear. Or notice and name all the objects that you can see
- Breathe in. Concentrate on your breathing. As you breathe out, say the number of the breath. Repeat this at least ten times.
- Splash water on your face.
- Sip a cool drink of water.
- If you are sitting, feel the chair under you and the weight of your body and legs pressing down onto it.
- If you are lying down, feel the contact between your head, your body, and your legs, as they touch the surface you are lying on. Starting from your head, notice how each part feels, all the way down to your feet, on the surface.

- Hold a cold can or bottle of soft drink in your hands. Notice the coldness and the wetness on your hands. When you drink it, notice the bubbles and taste.
- Hold a mug of tea or coffee in both hands and feel its warmth. Then take small sips and really notice the tastes with each mouthful.
- Wear an elastic band on your wrist and gently flick it gently, so that you feel it spring back on your wrist. Notice the sounds you can hear and the sensations you feel when you do that.
- Anything else that links you to the here and now.

The third way that can help with panic attacks is mindfulness. Mindfulness, after some practice, allows a person to dissociate from what's happening. They can recognize that they are beginning to feel agitated and they can accept that that's how they feel at the moment. That takes all the emotional power out of the incipient panic attack. They can also recognize that they feel that way at this very moment, but know that the feeling won't last forever, and will probably pass in a few seconds or so. (Some estimates have been quoted that urges, eg for a cigarette or a bar of chocolate, last no more than eight seconds.) Once the person panicking has become fully aware of what is triggering their emotional response (assuming that there is a specific trigger) and how the emotional and physical responses are connected, they are better able to control how they feel. Using mindfulness techniques does require a great deal of practice before it becomes easy for a person to use – particularly when they are starting to panic.

These techniques aren't a replacement for our standard way of working, but sometimes clients need something to get them through their everyday difficult experiences in the period before our techniques start to work.

References:

http://the-uterus.tumblr.com/post/120395827855/anxiety-attacks-arent-always-hyperventilating-and

http://www.puckermob.com/moblog/what-you-dont-know-about-panic-attacks-and-how-to-help

http://www.livingwell.org.au/well-being/grounding-exercises/

http://www.sweetescapeyoga.com/simple-stress-relief-wsquare-breathing/

http://blog.humangivens.com/2012/10/how-does-deep-breathing-make-you-feel.html

http://www.nhs.uk/conditions/stress-anxiety-depression/pages/understanding-panic-attacks.aspx

Fear of driving

A case study of the work done with one lady who was scared of driving.

Rachael, not her real name, booked an appointment to see me because she was feeling very unhappy when she had to drive down the motorway. By the time I first saw her, she said that she was getting panicky on small bendy roads as well. She also told me about last Friday, when she had driven up to the motorway, then round the roundabout and back along the main road. She'd stopped at the first lay-by and burst into tears – eventually phoning her husband to come and pick her up because she couldn't drive any further.

There are no figures available, but anecdotally, fear of driving on the motorway does not seem to be that uncommon a phenomenon, and many people that I've spoken to say they know someone like that. The fear, or phobia, most often remains invisible because the person suffering will simply avoid using the motorway and will drive along the ordinary roads instead – or, better still, get someone else to drive.

For Rachael, things were continuing to get worse. Not only was she now experiencing feelings of high anxiety on small roads where she was unable to see far ahead or behind, she was beginning to be reluctant to drive in the dark.

Rachael was scared that other drivers might do something dangerous, she was scared that something unknown would spontaneously happen to her car, she was scared that she would lose control of the car or that she wouldn't know what she was doing, and she was scared that she'd have a panic attack on the motorway.

> A panic attack has symptoms similar to doing strenuous exercise, eg fast breathing, sweating, raised heart rate, shaking. These are all the result of the adrenalin in your body. Sweaty palms make gripping easier. Vomiting and defeacating are way of putting of would-be predators.

Rachael's goal was to be able to go back to being able to simply drive from one place to another without giving it a second thought – to be able to drive pretty much on autopilot. And she wanted to *not* to wake up in the morning of a day when she had a motorway drive worrying about it.

From our conversation, it was clear that Rachael was stressed. Everyday events were filling up her metaphorical stress bucket, and every night, although she was sleeping, she wasn't emptying the stress bucket. She also wasn't relaxing enough. It seemed that she was always on the go and didn't have unstructured time to herself when she could just do anything she wanted.

Like most people with phobias, Rachael's primitive brain was trying to protect her. It knew that if she drove slowly she arrived safely. It knew that if there were no other cars around her, she would drive along safely. And it was exaggerating each of these actions to make her even more safe! The consequence was that she was driving really slowly on the motorway and even stopping on smaller roads to let other cars pass her.

I decided to come at the condition in a two-pronged approach. Because work was insisting that she drove between offices, she needed a strategy to be able to cope with the frequent journeys. So finding ways for her to cope was one thing. Secondly, I needed to find a way to remove the underlying issue (the real problem), so that, in future, she wouldn't need to use any strategies to successfully drive along the motorway or the smaller local roads.

Quick wins for calming down when you begin to feel stressed are to breathe deeply for a couple of breaths (taking care not hyperventilated!). Secondly, you can press on the vagus nerve on the chest or do neck extension exercises. These stimulate the parasympathetic part of the vagus nerve and can help make the body feel calmer.

I suggested that she drove along listening to music that she enjoyed and that made her feel calm. (Perhaps Elgar would be a better choice than heavy metal bands – but the choice depends on the person listening.) I suggested that it might take her mind of things if she were to sing along with the music.

Someone had told her to commentate on what she doing while driving. This led to her vocalizing her disquiet with the road and other drivers and eventually made her feel much worse. I suggested that she might sing. Or she could count backwards from 100. This is much like with a golfer who suffers from Yips. By getting the intellectual part of their mind involved in a task, it can't interfere with a well-practiced habit – in this case, driving. An alternative is to repeat a positive mantra, eg say, "I am an excellent driver", over-and-over again.

The alternative is to stop talking or singing altogether by touching the roof of your mouth with your tongue. This works by stopping people having that negative inner dialogue because people tend to move their tongue (even only a little bit) when they are criticizing themselves, and it physically can't happen when the tip of your tongue is pressed up against the roof of your mouth.

Another very useful technique for speedy relaxation is, what's called, 7-11 breathing. You teach the client to breathe in for a count of 7, and then breathe out for a count of 11. This works best using 'diaphragmatic breathing', which is where your diaphragm moves down and pushes your stomach out as you take in a breath. This is better than lung breathing.

This kind of breathing stimulates the parasympathetic nervous system (the rest and digest part) and so lowers emotional arousal. When Rachael breathed like this for five minutes or so, she found that she felt more relaxed. In addition, the fact that the intellectual brain was counting up to 7 and then 11 meant that it was kept busy and her good driving habits could do the job of getting her safely to her destination.

Lastly, we looked at how she could anchor the feeling of being relaxed and perfectly comfortable with what she was doing. Basically, the simple anchoring technique involves thinking about (or imagining) a time when you were completely relaxed. You have to remember (or imagine) the sights, sounds, and smells. You then fire an anchor. That means that you decide on an action that while being perfectly normal is not one you usually do. So, some people put their thumb and middle finger together. Others rub their earlobe, or their wrist (or whatever feels right). By repeating the thoughts and feelings, followed by the action, a person (in true Pavlovian style) associates them

I have a fairly standard procedure for a hypnotherapy session.

After the usual smiling and small talk as we enter the consulting room, I always use the standard, "what's been good about your week?" question. I always expect at least five things and I always try to drag out more. Why do I try for more? It's because I want them to spend the week between sessions noticing good things so that they can tell me when they sit down the following week. And that's because I simply want them to notice good things – which all too often people don't. They just take good things for granted. I also pause long enough for them to fill the gap in the conversation with another 'good thing'. And I tend to prompt them by asking, "what else?" With that, they are in the left prefrontal cortex of their brain, being positive and looking for solutions.

I always reward progress and exceptions to the problem, no matter how small, with a great deal of praise. And if they haven't made any progress or can't remember when things weren't equally bad, I tell them about a similar client and how they had managed to take that all-important next step (a proximity metaphor). I also like to tease out what strategies they used to help them make progress. This reinforces in their own mind that they do have the skills to not always be depressed (or whatever). And I tend to remind them of these on the couch, later.

I often ask, "What from the last session was particularly helpful?" This not only provides feedback for me, but also strengthens the positive effects of the previous session.

The next stage is scaling. "On a scale of 1 to 10 where 1 is really unhappy and 10 is jumping for joy, how happy are you?" I might also get them to scale other aspects of their life in order to show that they are making progress towards their goals. Once I have an answer, my follow-up question is always, "what would have to happen for you to be a (one more than the number they said)"? I give them a chance to answer, ie some thinking time. If they say that they don't know, I say, "suppose you did know", or, "pretend you do know". While this is an irritating question, it's surprising how often it produces a useful answer from the client.

together. So that after a few goes, simply performing the action (firing the anchor) makes the person experience the desired feelings – in this case, being very relaxed. So, when Rachael was driving along and beginning to get nervous, she could fire her anchor and relax again.

These all gave her something to do when she next had to drive – some tools that she could use in order to be able to make the journey without being a danger to herself or any other road user.

I used the first and second hypnotherapy sessions to help her to relax and to boost her confidence in her ability to make positive change and to regain her driving confidence. The next session was a rewind, where she took away the emotional impact of the fear. And the final session was a reframe. I asked her to tell me how she wanted to approach the drive between offices and I fed that back to her in the reframe.

> It's also a good idea to check that clients are still working towards the goal that they originally decided. I have had clients change their minds about what their goal is, others experience mission creep (where the number of goals increases), and some realize that their original goal is part of a much larger goal in their life. We will then discuss a solution-focused wording for the goal and I will write it down.
>
> It's usually useful to ask some of the SFBT questions at this stage – particularly the Miracle Question. The positive ideas generated by that, get clients in the right mood for getting on the couch.

By helping Rachael to empty her metaphorical stress bucket and helping her to deal with stress during the day (and so not fill her bucket quite so much), she was generally less stressed in all situations. That meant that her phobia didn't feel quite so bad, and the rewind and reframe helped make the phobia feel much less of a phobia.

That meant that she didn't fear the drive, and she didn't wake up in the morning worrying about the drive. And while she was doing the drive, nothing seemed too awful. And she managed to get onto the motorway. And she managed to overtake slower cars. And, although she wasn't totally relaxed at the thought of a motorway drive, she was able to do it. She acted 'as if' she could drive on the motorway and found that's what she was able to do. Over time, the slight fear she had receded into little more than a memory.

What to do when a client is 'stuck'

Here are some hints and tips on how to deal with a client who can't move forward with their life and doesn't know quite what to do next.

Most clients turn up because they want to lose weight, feel more able to give that presentation, or simply stop feeling so depressed. Getting them to agree a goal – something they want to move towards – is fairly straightforward. What can be much harder is when you have someone who feels completely stuck where they are and whose ultimate goal is not totally unrealistic, but is certainly a difficult one to achieve and one that is a long way away from their starting point – so far that they feel totally unable to take the first step. They are stuck – unable to move forward.

David McClelland came up with the Three Needs Theory in the 1960s, which suggested that people were motivated by their needs for achievement, power, and affiliation. According to this theory, my client had a high need for achievement, which meant he preferred working on tasks of moderate difficulty and avoided both high risk and low risk situations. He felt that trying to achieve his goal was too high risk, so he preferred to do nothing rather than risk the failure of not achieving it – he was stuck. But we're solution-focused hypnotherapists, so we don't worry too much about why people are how they are, we want to help them take steps towards achieving their goal.

In one of our early sessions, I asked him to describe his goal in great detail and used this in the hypnotherapy session so that he could picture himself successfully performing this new job. I suggested that he had a vision board at home so he could keep seeing things related to this goal. He didn't

> Asking a client to visualize the steps towards achieving a goal is a more successful way of helping people make changes than asking them to simply visualize the end goal being achieved.

like that idea because other people might see his goal and feel that he was failing to achieve it. I also asked him to imagine he had achieved his goal and was talking to himself now and explaining what steps he (this future self) had taken to achieve his goals – that way he could visualize what needed to be done. I then asked him, by what date he would have taken that first step. He laughed and couldn't be pressed into giving an answer. Being stuck was also being safe – or at least that's how it seemed.

As you can imagine, this all felt quite frustrating from my point of view. Sometimes, it can help a client to see a way forward if they simply state their goal in positive terms (what they will achieve rather than what they don't want to happen – eg I will be running my own company rather than I won't be working for my present company anymore!), and also to state by when they will have achieved it. I also like to walk them through what steps they need to take in order to get from where they are now to where they want to be. This 'chunking down' makes the whole process less daunting. For example, in order to get fit you need to join a gym. In order to join a gym, you need to search on the Internet for a local gym, and the client might agree to do this by Friday. The other advantage of walking through the steps is that the client can visualize themselves taking those steps. This, as well as setting a timeframe, identifies any resources that are needed – eg a need to buy running shoes. The last stage is to add sensory information to the picture. In trance, you can ask the client to see, hear, and feel what it's like when their goal is achieved. You can ask them about how they will behave when they have achieved their goals. And you can do the same with the steps along the way to achieving their goal.

NLP comes with the idea of 'ecology checking'. Sometimes a proposed change just doesn't feel right to the client. Although logically they want to achieve a change, their primitive brain makes it not feel right for them. When they look at the change in a disassociated way, they will see that it's just not right for them – something about it seems negative. In NLP terms, this new 'program' may conflict with existing 'programs'. It may result in the loss of benefits that are currently available. It may not address the presenting problem. It may be 'incongruent' within the client. The new program may actually create new problems. And there may be other benefits that could be achieved if a different program were used.

I tried using, what are called, Cartesian Coordinates on my client. The questions are:

- If you make this change, what will happen?
- If you make this change, what won't happen?
- If you don't make this change, what will happen?
- If you don't make this change, what won't happen?

Because my client was stuck, I was trying to make changes by making him think about things – to unstick him. Unfortunately, he didn't really answer the questions with anything I could work with. He seemed happier being stuck rather than being bold enough to make a change.

You're probably wondering how he responded to the Miracle Question. There were lots of "I don't know really" types of answer. I, of course, nodded encouragingly and didn't say anything. I wanted him to fill the void in the conversation. After a number of long pauses, he eventually decided that, when he woke up, he would feel happier and he would feel more confident about his ability to make changes in his life.

The early sessions were spent helping the client to empty his stress bucket and get back in control of his life. To help him to be better able to make proactive decisions rather than simply react to what was going on around him. In addition, he felt he had low self-esteem and we spent time helping with that. He said that he felt he was not as clever as the other people he worked with. He felt that people generally didn't like him as much as the others in his office. And he thought his opinions were of less value than anyone else's. He felt that he didn't deserve nice things. And all this was holding him back from making the changes he wanted.

As part of a couple of sessions, we looked at common cognitive distortions (taken from CBT – Cognitive Behavioural Therapy). These are exaggerated or irrational thought patterns that were originally identified by Aaron Beck. I've found that when people start to think about their thinking, they can start to change their thinking. Also, when they realize that these distortions are fairly common, they don't feel so bad about having used them themselves.

Here are some that we discussed:

- Always being right
- Blaming other people
- Disqualifying the positive – dismissing compliments
- Emotional reasoning – thinking something is true, solely based on a feeling.
- Fallacy of change – relying on social control to get cooperation from someone else.
- Filtering – only noticing bad things
- Jumping to conclusions – including mind reading and fortune-telling
- Labelling and mislabelling – a person's actions are due to their bad character
- Magnification and minimization – giving great weight to failure and little weight to success. It includes catastrophizing.
- Overgeneralization – inductive reasoning
- Personalization – taking the blame for everything that happens
- Should statements – expecting others to do what they morally should or ought to
- Splitting (all-or-nothing thinking or dichotomous reasoning) – seeing things in black or white.

One of the results of looking at this list is that people recognize that they are overgeneralizing. They realize that they are also spreading the bad news about themselves to others, which simply reinforces those feelings of low self-esteem. So, for example they may have problems finding a parking space today and so they proudly announce that they can never find a parking space. They order a takeaway meal and come home with one of the food items missing, so they tell everyone (and reinforce their belief) that they can't do anything right. Everything seems to reinforce their negative feelings (so that includes filtering and magnification and minimization from the list above).

So by helping people to recognize that they are being unfair in their view of themselves is the first step in helping them to increase their self-esteem. As a homework, you might ask them to write down every time someone says something nice – no matter how small a compliment that might be. Compliments might be simple thanks for a cup of tea, or compliment about a piece of work completed, or a work-related compliment such as, "you're always good at Excel". By keeping a note of all these small items of positive feedback, the client can see that their current view of themselves is one-sided.

Another technique that can help is to break through the all-or-nothing thinking – to show the client that there are shades of grey in between black at one end and white at the other. For example, some people may view the fact that they're not a size 8 as a clear sign that they are fat and therefore completely (remember this is all-or-nothing thinking) unattractive. The answer is to suggest that if a size 8 is at the top of the scale and a size 18 (or whatever) is at the bottom, where would you place a size 14 (or whatever size they are). This gets them scoring things in the 'grey' zone. You can do the same technique with anything else they do, like exercise or cooking. The important thing is to get them scoring and thinking in the 'grey' zone.

Other techniques that can help with all-or-nothing thinking are to ask:

- Just suppose you have all the money/support/time/energy/confidence/health you need. What other options/ways are there of looking at this?
- Think of someone you respect and admire. How might they look differently at this situation?
- What other possibilities/explanations/ideas are there that you haven't thought of yet?

Like so many things, low self-esteem is reinforced by the way people treat you – and the way people treat you depends on how you appear and how you behave. I often use the example of Matt Damon, the actor, who appeared in the *Bourne* series of movies as a hardened killer, whereas the same actor could appear as the naïve pickpocket in *Ocean's 11*. In both films it was the same person and in both films he was totally credible as the character he was playing. What I'm illustrating is that other people react to the way you behave. And they also respond to the way you appear. So if you look smart, they'll treat you like that (probably why people wear suits when they appear in court!). And if you appear slobish, then people will treat you with little respect. So, one way of raising self-esteem is to go shopping and smarten your appearance. Perhaps shave and have a haircut as well. People will react differently to your new appearance and behaviour and this will have a knock-on effect on improving your self-esteem.

Lastly, while in trance, I asked my client to visualize themselves feeling calm and relaxed, acting confidently, looking smart, appearing to be in control of the situation they are in, and recognizing the talents and skills they already have. The idea is to 'fake it until you make it'. You keep pretending to be something (relaxed and confident) until you find that that's exactly how you are.

It started to have the desired effect.

But one of the problems with being stuck is that you often feel that you can't do things. These are called 'limiting beliefs' and they can impoverish our lives. You hear people say things like, "I'm no good at maths", or "I can't cook", or "I'm not good enough to…". These are examples of limiting beliefs. To remove a limiting belief, it needs to be replaced with an empowering belief. See *Dealing with limiting beliefs*, p29, *Hypnofacts 3*.

Some people have an inner voice that is always being critical. My client didn't. If he had, I would have asked him to ignore it, or to challenge it by saying, "that's not what I really think". You can also take any characteristic and reframe it as a positive – see box (right). And you can ask your client to list their positive characteristics. If they were a superhero, what would their strengths be? If they review their strengths (intelligence, sense

'bad'	'good'
arrogant	confident
reckless	adventurous
can't sit still	exuberant
stubborn	single-minded
gullible	trusting
picky/fussy	meticulous
aggressive	assertive

of humour, determination, etc) each day, they'll soon come round to seeing themselves as better equipped to deal with life than they do now. And using hypnotherapy, I ask the client to visualize themselves performing the actions they wanted to perform, and do it using the strengths that they had listed.

At our next session, my client told me a good thing in his week where he had met some people who were associated with the industry he thought he wanted to get into. "Brilliant", I said, remembering how important it is to celebrate a client's successes and make a big deal out of it. Instead of smiling and high-fiving, he let his head fall and he looked at the carpet. "The trouble was", he confessed, "I felt rather shy the whole time and didn't talk much". I was surprised, I knew how much he wanted to meet these people and I hoped our work would have better equipped him.

I thought that it wouldn't do any harm to look at overcoming shyness along with our main work. We discussed ideas for conversation starters, like "Do you come here often? Did you get parked OK? How do you know our host? What made you choose this event?" My client soon got the idea and was coming up with much better examples. We also looked at body language – I asked him to model how a 'friendly' person would stand at a party and how he might have stood. We looked at facial expressions and eye contact. We talked about ways of joining a group and how you can make your face move to show you're reacting to what's being said before you start to talk. That way, everyone else thinks you are already part of the group. We talked about asking open questions and getting other people to talk about themselves – and how that will usually lead to them liking you a lot! And we looked at rapport building – how he could find links with other people (eg they both like football, or going to the cinema, etc). We discussed mirroring and matching their movements, their breathing rate, and even their type of language. On the couch, we rehearsed these techniques. We also looked at ways of calming down if he began to feel embarrassed. We talked about 7-11 breathing and imagining that he was relaxing in his favourite place and stepping into that feeling. And his homework was to join as many social 'groups' as possible to practice his techniques – so that, by repetition, the fear diminished (plus the bucket-emptying we were doing was going to help). And we kept discussing his goal and the benefits of acting as if he wasn't shy.

So now we could focus on bucket emptying and confidence. I continued to use useful paragraphs from our standard scripts (language patterns) when on the couch. And I also developed some of our earlier work – thinking about how a confident person would stand and behave at a party or conference. We talked about acting as if he was confident – the old 'fake it till you make it' idea. We imagined how it would feel to be ten times more confident than he was at the moment and then I asked him to step into that person's body and just feel what it was like. We then imagined what it felt like to be people he knew who were very confident and stepped into their bodies – letting their feelings wash over him (so he knew what being confident felt like). As homework, I asked him to write down good things that happened to him (no matter how small) and every time he acted in a confident way. Then I asked him to review his list before he went to bed each night, and first thing the following morning. And gradually he realized that he could be confident and good things did happen around him.

When we chatted in the first part of the session, we focused on the situations where he felt confident, and in the trance session we visualized him being confident in other situations. We also imagined how other people would see him (and talked about how little attention people really pay to other people). Like with people who want to lose weight or get fit, I asked him to spend more time with confident people, and less time with negative non-confident people. We also discussed the idea that there's no such thing as failure, only feedback – and how, if things didn't go right, it wasn't the end of the world, it was just useful information for next time. I used the Edison inventing the light bulb metaphor a lot. And we reframed social situations as 'experiments'. So after each experiment he could look at the results and decide what to do differently in the next experiment to get closer to his goal of being confident. And we continued to identify and celebrate every success.

I was really pleased with the improvements that were obvious in my client. He seemed to walk taller and he was going to meetings and social events, and reporting back that he'd felt comfortable and had contributed to conversations. In his Miracle Question answers, he was looking forward to a new job as part of his new career. I started talking about the longest journey beginning with a single step, and asked him what his first step towards this new career will be. I expected to hear all sorts of plans and ideas, but instead he almost visually wilted in his chair. "I don't know", he said. "if you did know, what would you do?", I asked. He smiled and said nothing. "if you had a friend in a similar situation, what would you advise?", I asked. He shrugged. I asked him what his choices might be, and again he hesitated and then said he wasn't sure – and yet I was sure that in an earlier session we'd chatted over some ideas. I was beginning to wonder whether he was anxious about making a life-changing decision, or whether he was anxious about making a decision and failing. We seemed almost back at the beginning again.

There's a phrase – paralysis by analysis – that can affect lots of people in lots of different situations. It's where they think about (brood on) what they can do to such an extent that they can't choose what's the best thing to do. I suggested that doing nothing was actually a choice. But he was stuck trying to make the right decision. I certainly wasn't going to validate any decision he made for him. I told him that people make better decisions (although I couldn't remember the source) when they need to go to the toilet. I suggested he find a time when he needed the loo, ask himself what he should do about his life, and write down the decision before going to the toilet. I also tried getting him to choose between different things as quickly as possible (like do you prefer fish and chips or a pizza?) – and then gradually worked towards harder questions (do you want to be in the same job this time next year or a different one?). Gradually, my client became more confident about making choices – realizing that most didn't matter, and getting more-and-more confident about making harder ones. He realized that you're never going to get all the information you need to make the 'right' decision. And, as you do get more information, it is simply confirming what you thought already. The idea is not to make emotional decisions (like sending that e-mail to your boss as soon as you get back from a night out at the pub), but to be able to make a decision once you have some of the information you need and it feels like the 'right' thing to do. He also stopped wanting others to validate his decision. And I re-iterated the idea that there's no such thing as failure only feedback. So when he tried something different at the Indian restaurant and didn't like it, it wasn't a wrong choice, he'd just added information to his personal database about what not to eat next time!

We began talking more about what my client wanted to do. We discussed how he felt about different aspects of his dream job. And we looked at the first steps he could take to get there. And then, I asked him what else he would like to do that played to his strengths and interests. He became very excited about the idea of what he could potentially do. In many ways, he had set himself a goal of his dream job some years ago, and, in the meantime, other things had interested him. He'd bought a camera and was serious about photography and Photoshop, for example. It suddenly occurred to him that perhaps his dream job wasn't his dream job any more, there were other things and he'd already taken the first steps towards improving his skills.

I felt we were coming towards the end of our time working together. I felt that my client was now feeling happier and confident about making changes to his life. I thought I'd try Robert Dilts' Disney Method (which is similar to de Bono's six thinking hats technique in some ways). The technique involves using different thinking styles – realist, dreamer, critic, and outsider – looking at the issue from each perspective. I put four cards on the floor with those words on them and asked my client to step on each one in turn and tell me his thinking. The idea behind this technique is to help people see how they feel about a decision and explore alternatives. Firstly he stood on outsider and described an analytical view of his choices. Then he stepped on dreamer and brainstormed a number of alternative choices he could make. Then he stepped on realist and evaluated each of the ideas that the dreamer had come up with. He then selected the best idea and came up with a plan. And then he stepped on the critic, and identified strengths, weaknesses, opportunities, and threats (SWOT) with the plan in order to improve it. So after that session, he had a working plan of what to do next.

My client was feeling confident and self-assured. He was looking to develop his career in a slightly different direction to his original idea. He felt more confident with people and with making decisions. He felt he had a roadmap of where he wanted to go and what he wanted to do next with his life. He was really happy with the work we had done together – and so was I.

References:

https://en.wikipedia.org/wiki/Disney_method

http://www.uncommonhelp.me/articles/category/confidence-and-self-esteem/

http://jeannineyoder.com/nlp-life-coaching-technique-4-questions-make-change-easy-life-coaching-clients/

http://grassrootsnlp.com/book/free-nlp-patterns/nlp-techniques-well-formed-outcome

http://grassrootsnlp.com/book/free-nlp-patterns/nlp-techniques-checking-ecology

http://grassrootsnlp.com/book/free-nlp-patterns/nlp-techniques-flexible-response

http://www.harleytherapy.co.uk/cognitive-distortions-cbt.htm#ixzz3xJkbxiJ4

https://en.wikipedia.org/wiki/Cognitive_distortion

Raising a client's self-esteem

Some ways to help clients increase their self-esteem and enjoy life more.

Let's start by defining self-esteem – so we know what we're talking about. A person's self-esteem is a reflection of their overall subjective emotional evaluation of their own worth. So, there's no logical formula for calculating a person's self-esteem, it depends on their beliefs about themselves and their emotions about their abilities.

So, how does an individual arrive at a view of their own self-worth? It seems that contingent self-esteem comes from external sources such as what others say, how successful that individual is at life events, what competences they have, and how good their relationships with others are. Non-contingent self-esteem is thought of as true, stable, and solid. It doesn't depend on external validation.

It doesn't really matter why a client has low self-esteem, what's important is the effect it's having on their life. Clients with low self-esteem tend to notice their mistakes and weaknesses, and tend to ignore their successes. They also take the blame for anything that goes wrong. And they dwell on their failures. They then tend to act, more-and-more, in a way that confirms their lack of ability and weaknesses. Clients will try to avoid social situations that they find challenging, including coming to see a therapist. They will, in fact, avoid any challenging or difficult situation – which will reinforce their feelings of low self-esteem. In addition, they may well develop negative habits, such as drinking too much alcohol, as a way of coping.

> Martin Ross suggested three self-esteem states. 'Shattered' is where a person does not regard themselves as valuable or lovable. With 'Vulnerable', a person has a positive self-image, but they are often nervous and regularly use defence mechanisms. A person with a 'Strong' self-esteem has a positive self-image and enough strength so that bad events (so-called 'anti-feats') do not lower their self-esteem.

Cognitive Behavioural Therapy (CBT) addresses the problem by getting people to examine their thinking and notice when they are thinking negative thoughts. They then get the person to look for alternative explanations to the conclusion they drew. And they get them to look for evidence that counters their thinking. So, for example, if they think no-one likes them because someone at work made a cup of tea and didn't offer them one, then the explanation might be that the person was going into a meeting and was making tea only for the people in that meeting (or some other plausible explanation). They could then list all the people who do like them. That way, they can start to turn off the negative thinking and not start to act as if no-one likes them.

Other ways that you can help a client raise their self-esteem include:

- Encourage them to do regular exercise, eat healthily, and get enough sleep.

- Help them to see how they can challenge their negative beliefs.

- Encourage them to recognize their positive qualities and the things they are good at and to recognize the things they have already achieved. Encourage them to do this quite frequently.

- Encourage them to take part in activities that they enjoy.
- Encourage them to accept their limitations and set realistic goals.
- Encourage them to spend time with positive, supportive people rather than negative people.
- Encourage them *not* to spend time comparing themself to other people, particularly on social media.
- Encourage them to relax – perhaps spending time doing a hobby.
- Encourage them to remember times when they have been happy, relaxed, and at peace.
- Encourage them to accept compliments when they're given.
- Encourage them to avoid all-or-nothing thinking.
- Encourage them to build positive relationships.
- Encourage them to be more assertive, or, to begin with, act in the same way that assertive people they know do.
- Encourage them to say 'no'.
- Encourage them to volunteer. If work is getting them down, then they may get the respect they need from volunteer work.

Low self-esteem in a client can often be associated with depression and a feeling of being stuck and not having the ability to risk making a change. Overcoming this issue can make a huge difference to the client achieving their goal. And our usual techniques of emptying their stress bucket and making them feel more confident work well.

References:

http://en.wikipedia.org/wiki/Self-esteem

http://www.mind.org.uk/information-support/types-of-mental-health-problems/self-esteem/

http://www.nhs.uk/livewell/mentalhealth/pages/dealingwithlowself-esteem.aspx

http://www.relate.org.uk/relationship-help/help-separation-and-divorce/separation-and-divorce-common-problems/low-self-esteem

http://www.getselfhelp.co.uk/esteem.htm

http://www.overcoming.co.uk/single.htm?ipg=8613

The complete guide to overcoming any fear ever!

Here's some advice to help clients who are experiencing a fear or phobia.

Fears are perfectly normal and one of the ways that a person stays alive. Being fearful of dark places means that our ancestors weren't eaten by predators. Being fearful of high places meant that they didn't fall to their death – they took care and survived. So fear is a good emotion to ensure a person's safety and survival. However, some people experience a persistent fear that is considered by others to be excessive and unreasonable. These are usually thought of as phobias. The phobic feelings start when that person gets close to a particular situation or object. And, because of the power of their imagination, the feelings can also occur when the person anticipates the approach of the situation or object.

What distinguishes a fear from a phobia is that people only become physically and/or psychologically impaired by a phobia. And people tend to organize their lives so that they avoid ever being in a position where they are likely to experience a phobic reaction.

> "Things do not change, we change."
> Henry David Thoreau

Although people tend to avoid their phobic situation, there may be times when they have to face it. They may be asked to give a presentation, or have to fly for work, or be inoculated against some disease – or whatever. That's when they are likely to come and see you.

The standard regime for dealing with this situation is four sessions – although there will be times when this needs to be modified to suit the needs of the client.

The first session is the usual initial consultation. This is where you build rapport with the client and find out their details – such as why they have come to see you and what their goal is from the sessions. For marketing purposes, you also need to know how they heard of you. It's important to listen to them and to show that you are listening – active listening reflects back to the client what they are saying. Next you'll need to tell them about what to expect from hypnosis and how natural it is, and then spend some time talking about the brain. I always show them some diagrams of the brain so they know what we are talking about and they can see how small the amygdala is in comparison to the rest of the brain. And I describe an amygdala hijack of their brain. Their homework is to stream the audio track (or listen to the CD) and think about what we've discussed.

Session 2 is all about helping them to relax and to be more confident. I teach them the peripheral vision relaxation technique and 7-11 breathing as ways of getting relaxed whenever they want to. The scripts I use are around confidence and change. For homework I get them to think about the first occurrence of their phobia and the worse occurrence. I need them to bring these two 'stories' to the next session. And they have to listen to the CD/audio stream every day.

Session 3 is mainly revision and then we use the rewind technique to remove the emotion associated with the phobic events – whether that's fear of giving presentations, or spiders, or needles, or whatever. Remember that the video they watch has to start in a safe place for them and end in a safe place. The homework is to come up with a story describing how they would like to behave when faced with whatever they are phobic about. This gets them thinking along positive (goal-focused) lines. They also have to listen to the CD/audio stream each day.

In session 4, I teach them the anchoring technique, where I anchor them to a feeling of relaxation and/or confidence. That way, anytime they need to, they can become calm and confident. It's a great tool to have. Then, the main part of the session is a reframe. Some phrases that I've found useful to include in a reframe include:

- You're pleased/fascinated/really interested/amazed ...
- Surprised at how quickly the time passed...
- You feel perfectly happy as you...
- So, in control...

Explaining to people about the right pre-frontal cortex can have powerful results. When they are brooding over their fear, they are strengthening the pathways linking the particular situation or object to their phobic response – making it even stronger, and making them feel even more anxious and their reaction even more phobic. Rule 1 is to stop brooding and suggest what people can think about instead. Remember, telling people to *not* think about something has the completely opposite effect. So, suggest that they count their blessings, do some brain training exercises, sing a song, or whatever alternative brain filling activity works for them.

> "They say that time changes things, but you actually have to change them yourself."
> Andy Warhol

Like writing their own reframe, visualizations are an incredibly powerful way of changing a client's thinking. But a word of caution: if you just say that you want them to think about how they'd like to be in a social situation or when receiving an injection etc, it's quite likely that their right pre-frontal cortex will hijack their thinking and they'll be back to thinking how awful it will be. One way round this is to get them relaxed before they start thinking about it. Get them to start visualizing the situation after they've listened to your CD/audio stream, or after a nice warm bath, or even when sitting comfortably in their favourite chair. Then get your client to visualize themselves acting confidently in the situation.

An alternative is to get them to imagine someone they know (who isn't anxious) to be in the situation. Get them to visualize what this other person is doing and how they are feeling (the sights, the sounds, the feelings). They can then (in their mind), step into this other person, and see how it feels to be them – and associate the other person's feelings with them being in the situation. By repeating these two visualization techniques, the association between the response and the particular situation or object becomes reduced – putting the client back in control of their life. It

> "Life is not a matter of chance... it is a matter of choice."
> Ka

also means that (although just in their mind) they are no longer avoiding their phobic situation.

And always make sure your clients are focusing on what they do want and never on what they *don't* want. When they make their own reality, you want it to be the positive successful version.

Because the client has visualized how they want to behave when facing the particular situation or object, it becomes easier for them to act as if they aren't scared. It seems the mind checks the body for clues about how it feels. So, if you're standing upright, looking up, and smiling, the brain thinks you must be feeling confident and relaxed – and so it reinforces those feelings. And if you are with other people, they will react to you as if you are confident and relaxed. It's an incredibly powerful feedback loop that can help the client overcome their phobia. Some researchers asked people to look at cartoons while holding a pencil with their lips. When they were forced into a smile, they found the cartoons funnier than when their lips were forced into a frown. And using that 7-11 breathing makes it seem like you are in control.

Another physical technique is to get the client to touch the roof of their mouth with their tongue when they are not talking. This works in most situations, not just social ones, because when people engage in an inner dialogue, they tend to move their tongues, even if only very slightly. So, by putting the tip of their tongue against the roof of their mouth, they can't have this inner conversation – and they can't listen to themselves saying negative things.

In many situations, the client with the phobia is concentrating on how they feel – they are focused inwards. One technique is to get the client to focus outwards. So, if they have a driving phobia, get them to regularly check and remember the colour of the car they can see in their rear-view mirror. If it's a social phobia or they're having an injection, get them to notice the colour of the walls, any pictures, or what people are wearing. I sometimes introduce it as playing Sherlock Holmes. The character would notice what people were wearing, or how they were walking, or how many steps there were, etc. I ask them to pretend to be the Conan Doyle character as a way of getting them to focus outwards.

If people do have a social phobia, I remind them that everyone else is totally focused on themselves and barely noticing them. I remind them of the experiment, where people were taking a maths test when someone wearing a 'Lady in Red' T-shirt came in late, made a fuss, and then left. Strangely, hardly anyone in the room could remember very much about the person. Unless you are a well-known personality, most people won't really take much notice of you. So don't worry whether they think you're stupid – they don't, they're too busy worrying about what everyone thinks of them.

And it's possible to rehearse situations with the client. If they were worried about a social situation, you could practice some nice open questions that they could ask: What weather! Have you travelled far? Did you get parked OK? I like your xxx, where did you get it? The avoidance of closed (yes/no) questions means that the other person will have to say a few sentences. Most people are pleased to do this because they don't want to stand in silence, they want to chat. And soon the conversation becomes perfectly normal. And, because your client is being Sherlock Holmes, they will remember some of the answers and will be able to bring them back into the conversation later. This will make them very popular because other people will feel that they have really been listened too.

When it comes to facing up to a phobia, clients are likely to envisage only two possible outcomes – success and failure. And if they feel they've failed, they are likely to add depression to their list of symptoms! The first thing is to show them that attempting any task is a continuum – the grey area between the black at one end and the white at the other. CBT describes this all-or-nothing thinking as one of its cognitive distortions. For most things in life, success is relative. If I were to run a marathon and complete it, that would be success for me. For someone else, completing it in under 4 hours would be success. And for other people, getting as close to two hours 10 minutes would be their definition of success. And so with overcoming most phobias. Being able to do it at all is the first line of success, and later, doing it more and more easily is the new success level. Your clients need to define what success looks like to them. And the very fact that they were prepared to make an attempt is a success in itself.

> Mindfulness can also provide a form of treatment for anxiety and stress

There's a saying in NLP that there's no such thing as failure, only feedback. It's useful to convince your clients that they didn't learn to ride a bicycle in one session, or learn to play the piano or drive, or anything else that they can now do quite easily. It's the same with overcoming their phobia. Their behaviour will get closer and closer to their desired behaviour with practice. There's no magic wand that can just make it happen.

The other thing is that they are not in control of everything in the world. What I mean by that is that they might have done everything right to be able to overcome their phobia, but events outside their control may prevent them from succeeding. It may be that today they might feel able to fly away on holiday, but because of the air traffic controllers' strike, no planes are taking off. When things don't work out too well, they need to be able to externalize and see what factors beyond their control stopped them today.

The three core human activities that provide a person with the ability to face most situations are to get enough sleep, eat sensibly, and get some exercise. Encourage your client to ensure those three things are happening in their life and they won't face the day cognitively impaired in anyway. They'll most likely be able to stay in the intellectual brain and face their phobia. These three help a person to deal with stress.

> NLP's 'circle of excellence' provides a way to deal with a fear or phobia. Picture a circle on the floor. In this circle there is a version of you, behaving the way you want to behave, ie confidently dealing with a situation or object that you are frightened of. When you step into this circle, you will feel exactly the same way. And you can step into the imaginary circle whenever you need to.

For many phobias, there are obvious triggers – for example the nurse will put a needle on the syringe and that will be the trigger for a needle phobia. For other fears, like fear of death or cancer, the triggers will not be so obvious. It can be useful to identify triggers for these feelings starting. That way, your client can nip the feelings in the bud before they go into a full-on phobic response. You can get them to think of alternatives – as mentioned above. You can also suggest that they wear an elastic band round their wrist and ping it every time they start to have those feelings or automatic thoughts. That way, they'll get more aware of them starting and will be better able to halt them in their tracks. And so they'll prevent their worrying thoughts. The elastic band acts as a pattern breaker for negative self-talk and helps the client to choose a new behaviour pattern.

I ask clients how anxious they are about facing their phobic situation, where 1 is completely relaxed and 10 is where they are about to run out of the room. I do this at each session and it's usually very easy to see the numbers going down – particularly after the rewind session.

For most people, the rewind and reframe are the most powerful tools you have in your toolkit to help them overcome their phobia. While you're working towards that with them, these other techniques can be useful because their job may mean that they have to face up to their phobia before you have had enough sessions. Reducing the stress and anxiety in your client's life (bucket emptying) will usually reduce the strength of the phobic response anyway. One CBT technique to try is gradual exposure to the object or situation the client has a phobia of. But most often this is unnecessary.

Other techniques that you can teach your client to use whenever they find they are beginning to feel nervous are:

- Anchoring – creating a link between a simple action (like rubbing their ear) and the feeling of being relaxed and in control.

- Peripheral vision relaxation – using the calming effect of the parasympathetic nervous system to help them relax

- Circle of excellence (see box above) – creating an imaginary area that has all the skills needed to be successful at a task. The client simply imagines stepping into the circle to feel fully equipped for the task ahead of them.

- 7-11 breathing – breathing in for the count of 7 and breathing out for the count of 11 also uses the parasympathetic nervous system to help calm down.

- Mindfulness urge surfing – a technique of recognizing that you feel a certain way at the moment, and this feeling may get worse for a short period of time, but after that, it will get better.

Armed with these techniques, you'll be able to help your clients overcome any fear.

IBS and hypnotherapy

A look at IBS and how hypnotherapy can help.

IBS (Irritable Bowel Syndrome) is a common, long-term problem. Different people show different symptoms and some people are affected more severely than others. The symptoms may last for a few days or for a few months and may be associated with eating certain foods or periods of stress. Estimates suggest that one in five people may experience IBS, which usually develops when people are in their twenties. Estimates suggest that twice as many women are affected as men.

There is no cure for IBS. And the good news for us is that the National Institute for Health and Care Excellence (NICE) recommends hypnotherapy as a treatment.

> NICE recommend that people living with IBS, who do not respond to pharmacological treatments after 12 months, consider a referral for psychological interventions, such as Cognitive Behavioural Therapy (CBT), hypnotherapy, and/ or psychological therapy.

The most common symptoms of IBS (according to http://www.nhs.uk) are:

- abdominal (stomach) pain and cramping, which may be relieved by going to the toilet
- a change in your bowel habits – such as diarrhoea, constipation, or sometimes both
- bloating and swelling of your stomach
- excessive wind (flatulence)
- occasionally experiencing an urgent need to go to the toilet
- a feeling that you have not fully emptied your bowels after going to the toilet
- passing mucus from your bottom.

Some IBS sufferers also experience:

- lethargy
- feeling sick
- backache
- bladder problems (such as needing to wake up to urinate at night, experiencing an urgent need to urinate, and difficulty fully emptying the bladder)
- pain during sex (dyspareunia)
- incontinence.

Because of the impact IBS has on a person, they may also have feelings of depression and anxiety.

The cause of IBS is unknown, although there are suggestions that it's related to problems with digestion and increased sensitivity of the gut. There are suggestions that food passes through the GI tract too quickly, causing diarrhoea. Or it passes through too slowly, causing constipation. Or that it doesn't pass through at all. Or it may be that the brain becomes oversensitive to messages from the gut, so mild indigestion feels like severe abdominal pain. And often a period of IBS can start after a stressful event. Other triggers for IBS include: alcohol, fizzy drinks, chocolate, caffeine-containing drinks, processed snacks (crisps and biscuits), and fatty or fried food.

Diagnosing IBS is difficult because there is no specific test. Often their doctor will exclude other causes first such as IBDs (Inflammatory Bowel Disease) like Crohn's or ulcerative colitis, which leave inflammatory markers in their blood tests.

> Cochrane looked at the research evidence and found that the studies provide some evidence suggesting that hypnotherapy might be effective in treating IBS symptoms including abdominal pain. However the results of these studies should be interpreted with caution due to poor study quality and small size.

As well as hypnotherapy, your client with IBS may try keeping a food diary to identify any foods that seem to trigger an episode. People with diarrhoea may try cutting down on the insoluble fibre (wholegrain bread, bran, cereals, and nuts and seeds). If they have constipation, they might try increasing the amount of soluble fibre they eat and the amount of water they drink. If your client has persistent or frequent bloating, they might try a low FODMAP (Fermentable Oligosaccharides, Disaccharides, Monosaccharides And Polyol) diet. FODMAP carbohydrates (fruits and vegetables, animal milk, wheat products, and beans) aren't easily broken down and absorbed by the gut. As a result, they start to ferment in the gut relatively quickly, and the gases released can lead to bloating. Many people say that exercise helps to relieve their symptoms of IBS. The exercise needs to be strenuous enough to increase your client's heart and breathing rates. You may find some IBS clients are taking anti-spasmodic drugs, some are on laxatives, others are prescribed antimotility medicines (for diarrhoea), and others may be using peppermint oil. Some people find taking probiotics regularly helps to relieve their symptoms of IBS. And some people will be taking antidepressants.

One of the main benefits of hypnotherapy for IBS is that it can help a client to relax, which, in turn, can help them to manage stress – to empty their stress bucket. It can also be used to help the client to visualize themselves coping and decreasing their sensitivity to messages from their gut. Hypnosis can also improve a client's general mental well-being, and provide psychological coping strategies for dealing with distressing symptoms, as well as help suppress thoughts and behaviours that increase the symptoms of IBS.

References:

http://www.cochrane.org/CD005110/IBD_hypnotherapy-treatment-by-hypnosis-for-the-treatment-of-irritable-bowel-syndrome

http://www.nhs.uk/conditions/Irritable-bowel-syndrome/Pages/Introduction.aspx

Hypnotherapy and fibromyalgia

A look at what fibromyalgia is and how hypnotherapy can help people who suffer from it.

You may be faced with a client who comes to you asking for help with their fibromyalgia. Hypnotherapy can be used to help a client deal with the pain and the fact that they will be experiencing that pain for a long period of time. But first, it's useful to know what fibromyalgia is.

Fibromyalgia (or fibromyalgia syndrome: FMS) is a long-term condition where a person experiences pain and a heightened pain response to pressure all over the body. Other symptoms include: extreme tiredness; muscle stiffness; difficulty sleeping; problems with memory and concentration (called fibro-fog); headaches; and IBS (Irritable Bowel Syndrome). Fibromyalgia is often associated with depression, anxiety, and post-traumatic stress disorder. Certainly, hypnotherapy is able to help with depression and anxiety. Some people will experience dizziness and clumsiness, feeling too hot or too cold, restless leg syndrome, and pins and needles (paraesthesia).

> Clients may say they have hyperalgesia, which means they are extremely sensitive to pain. Or they may have allodynia, which means they feel pain from something that shouldn't be painful (such as a very light touch).

No-one knows what causes fibromyalgia, but it's thought to involve a combination of genetics and environmental factors. The condition appears to be triggered by physical or emotional stress, such as: an injury or infection; giving birth; having an operation; the breakdown of a relationship; or the death of a loved one.

It seems that people with fibromyalgia have low levels of serotonin, noradrenalin, and dopamine in their brains. These are neurotransmitters that we're familiar with. It's also been suggested that a disturbed sleep pattern may be the cause of fibromyalgia, rather than a symptom.

It's not clear exactly how many people are affected by fibromyalgia, although research has suggested it could be a relatively common condition. Some estimates suggest nearly 1 in 20 people may be affected by fibromyalgia to some degree. Fibromyalgia affects far more women than men. People usually begin to experience symptoms between the ages of 30 and 50, but it can occur at any age. Fibromyalgia is a difficult condition to diagnose because there's no specific test. Treatments include antidepressants and painkillers as well as exercise programmes.

Hypnotherapy can help in many ways. Firstly, it can help to reduce stress by helping the client to empty their stress bucket. It can help clients to relax and it can give them techniques to allow them to relax when they aren't in the hypnotherapist's consulting room.

Hypnotherapy can help sufferers to deal with the pain. You can get your client to visualize a pain dial and turn it down, and notice that the pain feels less. When clients feel they have some control over the pain, they also feel they have more control over their life, which has a knock on effect of helping them to relax and making the fibromyalgia seem less important in their life. They also feel better able to manage the next episode of fibromyalgia that they experience. You are, in many ways, empowering your client.

Showing clients simple breathing techniques (like 7-11 breathing) can help them to relax.

Hypnotherapy can also be used to help with sleep problems.

As mentioned earlier, clients with fibromyalgia often report symptoms of depression and anxiety. Hypnotherapists are well used to dealing with these conditions and are able to help their client with them.

You can use visualization techniques with your client to help them picture themselves dealing with the condition in future in the manner of their choice. This will help reduce the level of the symptoms the client experiences in the future.

References:

http://www.nhs.uk/Conditions/Fibromyalgia/Pages/Introduction.aspx

https://en.wikipedia.org/wiki/Fibromyalgia

Laugh and the world laughs with you

Here's why laughter can be the best medicine.

We've all had sessions with clients, even those who say they're a bit depressed, that are a real hoot. Almost anything that's said is turned into a joke. I've even heard of other (not hypno) therapists at a clinic complaining about the laughter coming from the solution-focused hypnotherapist's room. But surely the laughter is nothing more than an example of rapport building. It doesn't actually do anybody any good, or does it?

Yes laughter is good in so many ways. For example, laughing relaxes your whole body. The act of laughing increases abdominal pressure and movements of the diaphragm. These movements massage the vagus nerve, causing it to send a signal telling the body to relax (using parasympathetic nerves). These movements of the diaphragm also act like a pump for your lymphatic circulation. This assists the lymphatic vessels in carrying fluid through your body and helps your lymph nodes to clean and filter this fluid, removing waste products, dead cells, and even unwanted microorganisms. An increased lymphatic flow improves your immune system. As well as decreasing the levels of stress hormone (cortisone), laughter increases the numbers of immune cells (lymphocytes) and infection-fighting cells (phagocytes), thus improving your resistance to disease and ability to fight infection. It also causes the body to release endorphins. These act as pain killers and promote an overall feeling of well-being.

There's evidence suggesting that a hearty laugh relieves physical tension and stress, leaving your muscles relaxed for up to 45 minutes afterwards. Also, when we laugh, we stretch muscles throughout our face and body. This results in our pulse and blood pressure going up, and we breathe faster, sending more oxygen to our tissues. This can increase a person's energy levels and make them more productive at work. Laughter also, apparently, causes the release of nitric acid, which helps dilate blood vessels, which, in turn, protects your heart.

Laughter has been shown to help hospital patients with a range of illnesses, making them better able to cope with their illness and their treatment. It's also very difficult to feel angry, anxious, or sad if we are laughing. Laughter helps us keep a positive, optimistic outlook when we're experiencing difficult situations, disappointments, and losses.

In terms of emotional intelligence, laughter helps us view situations in a more realistic and less threatening light – which is probably helping clients to escape negative thinking and view the world from their intellectual brain. This change of perspective can make us more empathic and better able to understand other people's points of view. It appears that laughter can strengthen relationships and enhance the way we work as a team. It seems to promote group bonding and help to defuse conflict situations. And it makes us more attractive to others! Laughing also helps to increase our resilience to stress and find meaning in our life. It can also help people create a positive outlook that can be applied to all aspect of their life – definitely lifting what could be described as a low mood.

It also seems that laughter can help us think 'outside of the box' and be more innovative and creative – coming up with ideas that will help us to achieve our goals. Not only do we become more emotionally aware, we also improve our memories. The hormone cortisone, which is produced in moments of stress, can damage the neurons in our hippocampus and can even shrink the size of our brain. So, laughter reduces the amount of cortisone and helps with memory. In fact, one study found that people who laughed were able to learn and recall information in almost half the time of those people who didn't laugh!

As well as helping us create endorphins, laughter affects our opioid system, and both of those are associated with stress-induced emotional eating. 10-15 minute of laughing burns 50 calories (according to a 2015 study conducted by Maciej Buchowski, a researcher from Vanderbilt University). So laughter helps with emotional eating problems. Laughter can also reduce blood sugar levels. There was a study of 19 people who ate a meal and then sat through a tedious lecture – after which they had their blood sugar levels measured. The next day, they ate the same meal and watched a comedy – and had lower blood sugar levels than the previous day.

So, if laughter came in tablet form, clients would be queueing up to buy it and take it. Laughter makes people feel better, work better, and relate to each other better. So, we should definitely be encouraging it in our consulting rooms.

References:

http://www.adam-eason.com/science-laughter-laughing-really-good-health/

http://www.helpguide.org/articles/emotional-health/laughter-is-the-best-medicine.htm

http://www.laughteronlineuniversity.com/laughter-immune-system/

Cortisone gets a bad press

A browse through the evidence of why the often-demonized cortisone might even be good for you.

We all know that a lot of people are coming to see us with stress-related problems, and these are linked to them having excess cortisone hanging around in their bloodstream. And, as such, cortisone is usually (metaphorically) thought of as the criminal mastermind behind any number of issues our clients are experiencing. But we all know that evolution doesn't give us anything that doesn't offer some kind of advantage.

Cortisone is produced in the cortex of the adrenal gland. It is released in response to stress and low blood-glucose levels. It increases blood sugar levels through gluconeogenesis (the generation in the liver of glucose from glycerol, some amino acids, and some other lipids), suppresses the immune system, aids in the metabolism of fat, protein, and carbohydrates, and decreases bone formation. Cortisone also prevents the release of substances in the body that cause inflammation. These are all useful things in a fight or flight situation.

If you want to reduce your cortisone levels, there are ways:

- Magnesium supplements decreases cortisone levels after aerobic exercise.
- Omega-3 fatty acids can slightly reduce cortisone release influenced by mental stress, although this depends on the dose.
- Music therapy can sometimes reduce cortisone levels.
- Massage therapy can reduce cortisone levels.
- Laughter and humour can lower cortisone levels.

Cortisone levels can be increased by:

- Anything that activates the HPA axis, like viral infections, trauma, or stressful events.
- Caffeine.
- Sleep deprivation.
- Prolonged aerobic exercise.
- Severe calorie restriction raises cortisone levels.

Let's take a look at that HPA (Hypothalamic–Pituitary–Adrenal) axis. The hypothalamus secretes vasopressin and Corticotropin-Releasing Hormone (CRH). These stimulate the pituitary gland to secrete of AdrenoCorticoTropic Hormone (ACTH). This causes the adrenal cortex to produce cortisone. The presence of cortisone acts on the hypothalamus and pituitary to suppress CRH and ACTH production (called a negative feedback cycle).

Adrenalin and noradrenalin are produced by the adrenal medulla after being stimulated by sympathetic nerves.

The problems that we have to deal with are with people who have cortisone left in their blood stream and seem to be anxious or edgy all the time. In addition, it's thought that the HPA axis is "involved in the neurobiology of mood disorders and functional illnesses, including anxiety disorder, bipolar disorder, insomnia, post-traumatic stress disorder, borderline personality disorder, ADHD, major depressive disorder, burnout, chronic fatigue syndrome, fibromyalgia, irritable bowel syndrome, and alcoholism". That means plenty of work for us.

But what happens if a person doesn't make cortisone – what happens if we remove this 'Napoleon of crime' in the body? The answer is a disease called Addison's disease.

The initial symptoms of Addison's disease can include:

- Lack of energy
- Lethargy
- Muscle weakness
- Low mood or irritability
- Loss of appetite and weight loss
- Increased thirst
- Frequent urination
- Craving for salty foods.

Later symptoms include:

- Low blood pressure when standing up, causing dizziness and fainting
- Nausea
- Vomiting
- Diarrhoea
- Abdominal, joint, or back pain
- Muscle cramps
- Chronic exhaustion, which may cause depression
- Brownish discolouration of the skin, lips, and gums
- A reduced libido, particularly in women.

So, not having any cortisone in the body can be terminal, or at least lead to a very unpleasant experience. In fact, looking at all the positive things it does in the body, you could conclude that cortisone is actually good for you.

Our work as hypnotherapists is to help people to feel more relaxed and so produce less cortisone, and encourage them to take some exercise to naturally reduce the amount of cortisone in their body. And this can help with conditions as different as IBS, PTSD, and insomnia.

References:

Spencer RL, Hutchison KE (1999). "Alcohol, aging, and the stress response". Alcohol Research & Health 23 (4): 272–83.

https://en.wikipedia.org/wiki/Hypothalamic%E2%80%93pituitary%E2%80%93adrenal_axis

https://en.wikipedia.org/wiki/Cortisol

http://www.nhs.uk/conditions/Addisons-disease/pages/introduction.aspx

http://www.nhs.uk/Conditions/Addisons-disease/Pages/Symptoms.aspx

Hardwiring

Let's take a look at the vagus nerve and how it affects our behaviour, emotions, and nervous system.

We often hear people say that they feel something in their gut, or that something is heartfelt, and we probably take those expressions to be metaphorical and not literal. After all, the heart is just a pump for getting blood round the body and the gut is simply a place for digesting food, aren't they?

Strangely enough, there is a main nerve that wanders round the body linking the heart and intestines, as well as the throat and the lines in the corner of the eyes to the brain. It's the vagus nerve.

> Vagus means wanderer – it has the same root as the word 'vagabond'.

We tend to think that the brain is the most important part of the body when it comes to deciding behaviour. After all, the brain comprises only 2% of the body's weight, yet it receives 15% of the cardiac output, 20% of the total amount of oxygen consumed by the body, and 25% of all the glucose used by the body. But suppose the vagus nerve decided behaviour as well.

The Central Nervous System (CNS) comprises the brain and spinal cord. The Peripheral Nervous System (PNS) connects to the CNS and is divided into the sensory nervous systems, the somatic nervous system, and the autonomic nervous system. The sensory nervous system transmits signals from our senses to the spinal cord and brain. The somatic nervous system is under voluntary control, and transmits signals from the brain to our muscles. The autonomic nervous system works unconsciously and regulates bodily functions such as the heart rate, digestion, respiratory rate, pupillary response, urination, and sexual arousal. It is regulated by the hypothalamus. The autonomic nervous system has two branches: the sympathetic nervous system and the parasympathetic nervous system. The sympathetic nervous system is responsible for 'fight or flight' activities. The parasympathetic nervous system is responsible for the 'rest and digest' activities. Lastly, there's the Enteric Nervous System (ENS), which consists of a mesh-like system of neurons that governs the function of the gastrointestinal system.

Figure 1 illustrates many of the organs that the much-branching vagus nerve goes to.

Dr Stephen Porges in 1995 came up with the idea that the vagus nerve was more complicated than we first thought – and he came up with his Polyvagal Theory. The first, the unmyelinated vagus, originates in the Dorsal Motor Nucleus (DMNX), and has efferent fibres that regulate smooth muscles and heart muscle (but not skeletal muscles). It regulates peristalsis of the gut and sweating, and connects to the lungs, diaphragm, and stomach. It's responsible for heart rate, dilation of blood vessels, and blood pressure.

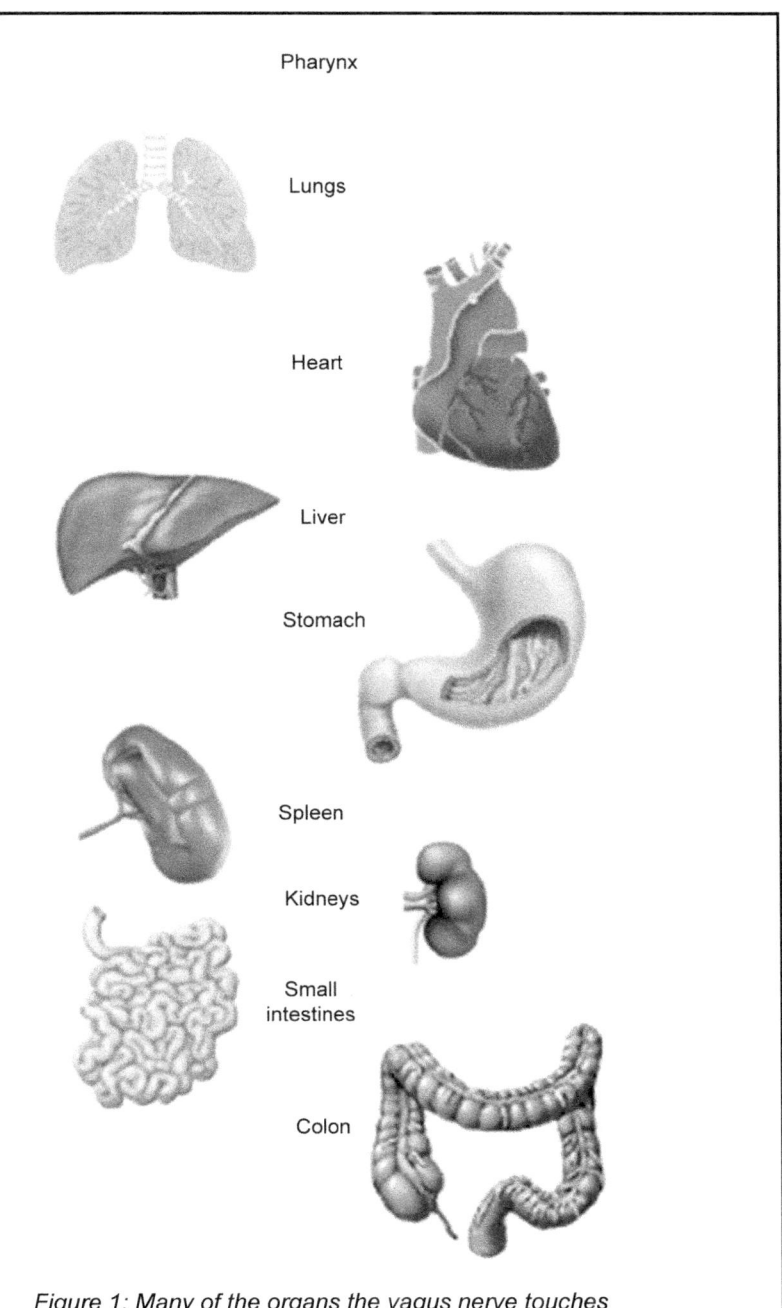

Figure 1: Many of the organs the vagus nerve touches

The second, the myelinated or mammalian vagus, originates in the medullary source of the Nucleus Ambiguus (NA), and its efferent fibres regulate the somatic muscles of speech and eating, as well as the larynx, pharynx, and oesophagus. It's associated with processes such as attention, motion, emotion, and communication. It provides a cardiorespiratory rhythm.

Porges suggests that some medical conditions are a result of competition between the two fibres because they have opposite effects on the same organ. The vagus nerve has afferent and efferent parasympathetic fibres allowing both motor and sensory parasympathetic action. It's estimated that 20 percent of the fibres control the organs (heart, digestion, breathing, and endocrine glands), and 80 percent transmit information from the gut to the brain.

> The suggested benefits of vagus nerve stimulation are:
> - It reduces inflammation.
> - It helps create new brain cells.
> - It lifts mood.
> - It improves memory.
> - It increases immunity.
> - It increases the level of endorphins, which reduces the sensation of pain.

> One way of fighting inflammation in the body is by engaging the vagus nerve and improving 'vagal tone'. Healthy vagal tone is indicated by a slight increase in heart rate when a person inhales, and a slight decrease when they exhale. Deep diaphragmatic breathing and slowly exhaling stimulates the vagus nerve and slows a person's heart rate and lowers their blood pressure. Clinical trials have shown that stimulating the vagus nerve with a small implanted device significantly reduces inflammation and improves outcomes for rheumatoid arthritis sufferers by inhibiting cytokine production.

According to the polyvagal theory, our nervous system has more than one defence strategy, and they are all outside our conscious awareness. The first to evolve (and used principally by amphibians and reptiles) is the freeze response. The unmyelinated vagus responds to threats by depressing metabolic activity. We use it when we experience a severe injury, it shuts us down and turns off our registration of pain. The second is the fight or flight response. This depends on the functioning of the sympathetic nervous system increasing metabolic output. The third stage is communication, and uses the mammalian myelinated vagus. It can rapidly regulate cardiac output to calm you down, and is associated with cranial nerves that regulate sociability through facial expressions and vocalization. The newer (in evolutionary terms) responses inhibit the older ones. People use the newest circuit to promote calm states, to self-soothe, and engage. When the parasympathetic nervous system is dominant, higher order cognitive functioning is possible, which enables a wider and more flexible range of behaviour. If that doesn't work, they can use the sympathetic-adrenal system for flight and flight. When the sympathetic nervous system is dominant, social behaviour is limited to survival strategies such as aggression, avoidance, or withdrawal. And if that's not possible or not working, the old vagal system can be used to freeze or shut us down.

So how does the nervous system know to try to communicate (social engagement) rather than fight or flight? It needs to assess the risk and then inhibit the more primitive behaviour. The nervous system processes sensory information from the environment and viscera without any conscious awareness (although it may involve subcortical limbic structures). This is called 'neuroception' to emphasize that it's a neural process, not a perception. Social engagement connects the social muscles of the face (eyes, mouth, and middle ear) with the heart. We use this to clear up misunderstandings, get help, ask for forgiveness, etc.

According to A D (Bud) Craig, emotions come from feelings in our organs and gut, and these feelings are sent using the vagus nerve to the Anterior Insular Cortex (AIC) in the brain. The AIC stores these feelings as a series of snapshots, and this makes up our working emotional memory. These feelings

> Vasovagal syncope occurs when a person faints because their body overreacted to a particular trigger, eg the sight of blood or extreme emotional distress. The trigger causes their heart rate and blood pressure to drop suddenly, leading to reduced blood flow to their brain, causing them to briefly lose consciousness.

are integrated with the social exchange to give us an emotional response to what's happening around us as well as a safety strategy. So, if our clients are feeling anxious with everyday life, perhaps the solution is in their vagus nerve!

We've said that the vagus nerve can help a person cool down after a stressful experience. Taking a long deep breath will stimulate the vagus nerve to calm down the body. Stimulation of the vagus nerve helps people with chronic depression who aren't responding to other treatments. Vagus Nerve Stimulation (VNS) can help with weight loss because it makes people feel full. VNS had positive results when treating anxiety disorders, Alzheimer's disease, and fibromyalgia.

So how can you stimulate your vagus nerve? Place your hand over the centre of your chest or even just take a lung full of air. You can constrict the pharynx by breathing in and out through your nose. Roll back the eyes (stimulate deep thought at the same time). Put pressure from the tip of the tongue against either the floor or the roof. Tug your ear. And finally squeeze those pelvic floor muscles.

I think this provides useful extra information for us to use in the first part of a session.

References:

Porges, S. W. (2008, February). The Polyvagal Perspective. NIH Public Access, PMC1868418

Craig, A. D. (2009). Emotional moments across time: A possible neural basis for time perception in the anterior insula. Philosophical Transactions of the Royal Society of London. 364,1933-1942.

http://attachmentdisorderhealing.com/porges-polyvagal/

http://biologyofkundalini.com/article.php?story=PolyvagalTheory

http://www.ncbi.nlm.nih.gov/pmc/articles/PMC1868418/

http://www.sott.net/article/256043-Polyvagal-Theory-Sensory-Challenge-and-Gut-Emotions

http://www.redpilllife.com/blog/holy-vagus-this-nerve-does-what/

A funny tummy and my brain!

A look at the link between your GI tract and how you feel.

We're all completely familiar with our GastroIntestinal tract (GI tract) or, as we familiarly call it, our gut. We know that it's responsible for digesting and transporting foodstuffs, absorbing nutrients, and getting rid of waste. But what few of us are aware of is how much it affects our brain and how big an impact it has on how we feel.

The current theory is that animals developed brains in order to coordinate their movements – to make them less random. And, once they had a brain, they started to use it for other things, like thinking and feeling (although they probably didn't evolve in that order).

It's interesting to note that the gut has a huge number of nerves and these nerves are different from other nerves in the body. The Enteric Nervous System (ENS) is one of the main divisions of the nervous system and consists of a mesh-like system of neurons that governs the function of the gastrointestinal system. Some people refer to the gut network of nerves as the 'gut brain'. Clearly, with so many nerves, the gut must do more than simply digest food. And we find that many of our common phrases and sayings refer to 'gut feelings' or actions.

> Stimulating the vagus nerve can make people feel comfortable or anxious – depending on the frequency used. In 2010 the European Union approved stimulation of the vagus nerve as a treatment for depression in humans.

Signals from the gut go to different parts of the brain, including the insula, the limbic system (including the amygdala and hippocampus), the anterior cingulate, and the prefrontal cortex.

What evidence is there that our gut affects our brain? Firstly, there's evidence from an experiment on mice. Mice were put in a container of water and forced to swim around (to find dry land). Depressed mice swim for less time before they give up. Inhibitory signals are transmitted more efficiently in their brains than motivational or driving impulses. Antidepressants are tested on mice because, if the mice tend to swim longer, the antidepressant must be working. But what happens if you feed your mice with bacteria that are good for the gut? That's exactly what they did, and the mice swam for longer and their blood contained fewer stress hormones. The mice also performed better in memory and learning tests too. But once the vagus nerve (it links the brain and the gut) was cut, the mice showed no improvement.

Another example is when mice are given antibiotics, which affect the gut biome. This has been found to lower levels of BDNF (Brain-Derived Neurotrophic Factor), a protein in the brain used in creating new neurons.

Because of the large surface area of the gut, it is our largest sensory organ, and information is sent to the brain so it can know about what's going on in this internal environment. In fact, for a baby, most of what it knows about the world comes from its gut – and that affects how it feels. A baby cries if it's hungry or needs its nappy changed. It smiles with the pleasure of a full stomach. This link between the gut and the brain doesn't disappear as we grow older, and it can still affect our mood and sense of well-being. In fact, a 2013 study found that after four weeks of swallowing certain bacteria, there were unmistakable alterations to areas of the brain associated with pain and emotions of people.

In another experiment, people had small balloons in their intestines inflated. Healthy patients didn't show any unusual brain activity. People with IBS (Irritable Bowel Syndrome) showed activity in the emotional centre of the brain, making them feel uneasy. IBS sufferers are known to experience a higher than average incidence of depression and anxiety. Similarly, people who suffer from Crohn's disease or ulcerative colitis also have increased rates of depression and anxiety.

Let's suppose that you're under pressure at work – you have to complete some long and complicated task by Friday. You are feeling stressed. Your brain needs more resources (food and oxygen) in order to complete the work, and it gets it by redirecting blood that would have gone to the gut towards the brain. Messages travel through sympathetic nerve fibres to use less energy in digestion and produce less mucus as well. If the stress continues, and supplies to the gut stay at the lower level, the consequence can be fatigue, loss of appetite, general malaise, and diarrhoea. If the stress continues even longer, the health of gut will decrease, resulting in a weaker gut wall. As a result of this, immune cells in the gut (and there are more here than anywhere else in the body) become more sensitive – and so you become more sensitive to what you eat. This suggests that many food allergies are a consequence of stress.

> Probiotics are live bacteria and yeasts that are good for your health, especially your gut.
>
> Prebiotics are substances that induce the growth or activity of bacteria and yeasts in your gut.

Another suggested consequence of prolonged stress is that it affects the bacteria that live in your gut – making it a better environment for some and worse for others. Remember what happened to the mice when the bacteria levels changed? The same can happen to you. And there is a time delay after the stressful period ends before the gut bacteria return to their original levels.

It may be that the brain remembers the negative feelings from the gut and is less likely to put itself in the same stressful situation again. It may be why people feel less keen on giving a presentation, even though the first one appeared to be quite successful.

Going back to mice, we find that some strains are more timid and docile while others are more adventurous. What do you think would happen to the behaviour of the mice if, somehow, their gut bacteria were swapped around? By using antibiotics, that's exactly what scientists achieved. They killed all the existing gut bacteria and then 'rebooted' them with the bacteria usually found in the other type of mice. The result was the 'shy' mice became more gregarious, and the 'outgoing' mice became more timid. There's no saying that the same thing would happen in humans, but it does add weight to the argument that the gut does affect the brain.

It's recommended that meal times are stress free events because any kind of stress inhibits digestion, which means we get less energy from our food and it takes longer to digest it, which adds to our stress.

We know that travel sickness tablets numb the nerves of the gut. It's also been found that as the feelings of nausea disappear, so do any feelings of anxiety. Alcohol reaches the gut before it reaches the brain. Perhaps its relaxing effects come from what it does to the nerves of the gut? And if you want to achieve the same effect with bacteria, Lactobacillus reuteri is able to inhibit the pain sensors in the gut. Also, Lactobacillus plantarum and Bifidobacterium infantis help with IBS.

The good news for us is that hypnotherapy has been proven to work with patients with IBS and is a recommended treatment by NICE (The National Institute for Health and Care Excellence).

If 95 percent of all the serotonin produced by the body is in the gut, it can come as no surprise that taking SSRI (Selective Serotonin Reuptake Inhibitors) has an impact on the gut as well as an effect on the brain. It has been suggested that the next breakthrough drug for depression will only affect the gut and not the brain. It's an interesting idea.

Your gut contains three types of bacteria. Bacteroides produce enzymes for digesting just about any carbohydrate. It may be these bacteria that produce more usable simple sugars from the food we eat and allow some of us to absorb more calories than someone else eating the same food with fewer of this family of bacteria in their gut. Prevotella tend to dominate in the gut of vegetarians. The third group comprises any other bacteria that can be found in the gut.

> Invasive bacteria can change the way people behave. Toxoplamata in the gut can cause animals and humans to seek out dangerous activities that they wouldn't otherwise. For example, people may self-harm.

It may be that the clients we see for anxiety and depression, or even fear of giving another presentation, may make faster recoveries by changing their diet. Clearly they need to reduce the stress in their life and empty the stress bucket, which we can help with. But, in many ways, it's a feedback loop. Any changes in the brain cause changes in the gut. And any changes in the gut can cause changes in the brain. The fact that you have a funny tummy today could be caused by stress at work yesterday and could cause your bout of depression tomorrow.

References:

Giulia Enders; Gut; Scribe Publications; ISBN-13: 978-1922247964

Flegr J: 'Influence of latent toxoplasma infection on human personality, physiology, and, morphology: pros and cons of the toxoplasma-human model in studying the manipulation hypothesis'. In: J Exp Biol. 2013 January 1: 216 (pt 1): pp 127-33.

Slattery DA1, Cryan JF. 'Using the rat forced swim test to assess antidepressant-like activity in rodents'. Nat Protoc. 2012 May 3;7(6):1009-14. doi: 10.1038/nprot.2012.044.

Bravo JA. 'Ingestion of Lactobacillus strain regulates emotional behaviour and central

GABA receptor expression in a mouse via the vagus nerve'. In: Proc Natl Acad Sci USA 2011 September20: 108 (38) pp 16050-55

Enck P et al. 'Therapy options in Irritable Bowel Syndrome'. In: Eur J Gastroenterol Hepatol 2010 December: 22 (12) pp 1402-11

http://yourbrainhealth.com.au/gut-brain-health-paradigm-shift-neuroscience/

Diet and depression

Here's a review of recent research findings showing that what you eat can affect how you feel.

Depression is characterized by weeks of low or sad mood, diminished interest in things that used to be pleasurable, weight gain or loss, inappropriate guilt, fatigue, difficulty concentrating, and recurring thoughts of death. It's estimated that 7 percent (roughly sixteen million) Americans have at least one depressive episode each year.

A 2008 study found that happier patients were healthier – but which came first. A follow up study made people ill (they actually dripped the common cold virus into their noses). The study found that one in three of the people initially rated as anxious, hostile, and depressed went down with a cold, but only one in five of the people rated happy and relaxed. So mental health does affect physical health.

If I don't want to be depressed, what foods should I eat? And why?

The first thing to do is cut down on the amount of meat that you eat. Animal products contain the proinflamatory product, arachidonic acid, which can adversely impact mental health through a cascade of neuroinflammation. Ibuprofen works by blocking the conversion of arachidonic acid into inflammatory end products. Your body makes all the arachidonic acid it needs – so doesn't need any more from your diet. Chicken and eggs are the biggest source of arachidonic acid, followed by beef, pork, and fish.

Researchers took a group of people who ate meat for at least one meal a day and put them on a diet without meat or eggs. After two weeks, the subjects experienced significant improvement in the measures of their mood. In another study, overweight and diabetic staff at an insurance company were encouraged to eat a whole-food plant-based diet. After five months, compared to a control group that could eat what they wanted, the people had increased energy, better sleep, and significant improvements in their general health and mental health.

Nutritional neuroscience found that a higher consumption of vegetables may cut the odds of developing depression by 62 percent! If you assume that the enzyme monoamine oxidase (MAO) breaks down excessive monoamines in the brain – including our favourite neurotransmitters serotonin and dopamine – then decreasing the amount of MAO will increase the amount of these neurotransmitters available. Apples, berries, grapes, onions, and green tea, as well as cloves, oregano, cinnamon, and nutmeg contain phytonutrients that can reduce the effect of MAO, which could make you happier.

There's also evidence from the USA that the more coffee you drink, the less likely you are to commit suicide! People who drink two cups a day have half the suicide risk of non-coffee drinkers. People who drank six cups of coffee a day are 80 percent less likely to commit suicide. However, drinking eight cups a day goes with an increased risk of suicide.

Aspartame (the sweetener) seems to be something else to avoid. A 1980 study found that people with mood disorders were particularly sensitive to artificial sweeteners. A more recent study found that after eight days on a high aspartame diet (around half of the daily intake recommended by the FDA) participants exhibited more depression and irritability, and performed worse on tests.

> Antioxidants react with free radicals, which are produced by chemical reactions in the body. Free radicals are associated with ageing, cancers, strokes, and cataract, and can damage the DNA in cells.
>
> Flavonoids, which are found in plants, and some vitamins and some minerals are antioxidants.

A study of nearly 300,000 Canadians found that greater fruit and vegetable consumption was associated with lower risk of depression, psychological distress, mood and anxiety disorders, and poor perceived mental health. The researchers suggested that eating antioxidant-rich plant foods may dampen the detrimental effects of oxidative stress on mental health.

Lycopene (the red pigment in tomatoes) has a very high antioxidant effect. A study of 1000 elderly men found those who ate tomatoes or tomato products daily had just half the chance of becoming depressed compared to those who ate tomatoes once a week or less.

Folate (Vitamin B9) is found in spinach and has an impact on depression. A low dietary intake of folate can increase the risk of severe depression threefold. Studies show that eating an actual plant containing folate is effective, whereas simply eating a supplement containing B9 isn't.

> It's been suggested that depression may be a result of inflammation. Here are the top 15 anti-inflammatory foods you can add to your diet: green leafy vegetable, bok choy, celery, beets, broccoli, blueberries, pineapple, salmon, bone broth, walnuts, chia seeds, flaxseeds, turmeric, ginger.

It's unlikely that any of our meat eating clients are going to stop eating meat completely and move to a vegan diet. But at every meal they have a choice, and if they don't want to continue feeling depressed, maybe they can be encouraged to eat a little less meat and a little more spinach, tomatoes, apples, berries, and grapes; and perhaps they can wash it down with a green tea rather than a diet fizzy drink.

References:

Dr M Greger, G Stone; How Not To Die: Discover the foods scientifically proven to prevent and reverse disease. Macmillan; 978-1447282440

The primitive brain gets a bad press

A look at why the limbic system – part of the brain – is so useful, even though it is often not well regarded.

There are some hypnotherapists who think we walk around on a knife edge – sometimes we fall one way and we're in the intellectual brain and all is right with world, and other times we drop into the pit that is the primitive brain and that way lies ruin! It's clearly not true – so what is the true picture?

The truth is that different parts of our brain evolved over time to help us function better in the environment in which we lived. Our brainstem is the most primitive part of our brain and that keeps all our important bits working without us needing to think about it – things like our heart beating, our lungs breathing, and our GI tract squeezing food through it (peristalsis). Our primitive brain – the limbic system – evolved in mammals and gave us a huge evolutionary advantage. We could build memories in the hippocampus. We could allow lots more sensory impulses to be received because we could now ignore so many of them (delete, distort, and generalize) in the thalamus. We had a better way of regulating everything going on in the body (homeostasis) in the hypothalamus. And we could feel things – experience emotions, particularly fear – which increased our chances of survival. They made us better at the 4Fs (fighting, fleeing, feeding, and reproducing!). Lastly, the great apes (and whales and dolphins) evolved the intellectual brain, a way of recognizing relationships in society and making logical decisions, problem solving, maintaining attention, and controlling those primitive emotions. It also gave us a way to analyse what had happened so we could do things better next time (although some people just use it to brood!).

The other vital piece of information that we need to know is how much energy the brain uses. It uses about a fifth of all the oxygen used by the body. And that is a lot. So our highly-efficient body tries to reduce the energy demands of the brain. How can it do that? Well it turns off as much thinking as possible!

Although people do use every part of their brain (otherwise why would we have evolved the other 90 percent if we really did use only 10 percent of it!), energy efficiencies mean that you try to use as little as possible at any one time. You can safely drive home without remembering anything about the journey. You can do lots of things without using any more than the barest minimum amount of brain energy. Think of the brain as the 'greenest' organ (in terms of energy conservation) in your body.

You create habits very quickly, and then you repeat those habits – it saves thinking which shoe to put on first, for example. Your hand reaches out for the alarm clock without you needing to think about its location (if you always put it in the same place). The brain's coordination centre for habits is the striatum. It is connected to the prefrontal cortex and the midbrain. The midbrain provides input from dopamine-containing neurons. (Dopamine is the neurotransmitter associated with creating positive feelings related to reward and events of emotional significance.) Once a habit is stored, the infralimbic cortex causes a person to carry out the habit when they are triggered by a particular cue, situation, or event.

If I were to gently throw a ball towards you now, you would probably be able to catch it. And it wouldn't be your intellectual brain calculating the parabolic flight of the ball that

> The neurotransmitters, dopamine, noradrenalin, and serotonin are only made in the synapses of the primitive brain.
>
> Neurotransmitters pass electrical impulses across synapses from one neuron to another. So how can they have any impact on the rest of the brain (almost as if they were hormones)? How could they make you happy (serotonin) or more motivated (dopamine)? The answer is neuromodulation.
>
> Neuromodulation is where a neuron uses one or more chemicals to regulate diverse populations of neurons. Neuromodulators (ie dopamine, serotonin, noradrenalin, acetylcholine, and histamine) are secreted by a small group of neurons, and diffuse through large areas of the nervous system, affecting multiple neurons.

helped you, it would be ball-catching rule-of-thumb (heuristic) that you developed when you were a child. And there are loads of these heuristics that we use all the time to make life easier (and use up less oxygen and sugar in the brain).

All these examples are people using their primitive brains. Those pesky primitive brains aren't there just to protect you at the first sight of a polar bear – they're working for you pretty much all the time. It's only when you have to make a hard decision that you use all that extra energy in your intellectual brain.

In fact, using your primitive brain can be better than using your intellectual brain! Have you ever had a client with a sporting problem. Perhaps they have yips or dartitis, or any number of similar conditions. The problem for them is that their primitive brain is perfectly capable of throwing the dart accurately, or making that golf swing, but the intellectual brain keeps cutting in and wants to check everything is OK. It starts to think about the best way to swing that club, which prevents your primitive brain accessing those perfectly honed habits and heuristics. That's why we need to get them to sing, or fill up their thinking brain with something else, so the primitive brain can get on with doing its job of playing the sport.

Similarly, have you ever tried to teach a family member how to drive? Your primitive brain is such an expert driver that it moves your hands and your feet without you needing to consciously think about it. And that's where the trouble starts. You're learner driver can't plug their primitive brain into yours and download the information. They need you to tell them what to do. You need to verbalize something you haven't needed to verbalize for a very long time. It's hard and it's slow, and by the time you realize that you've missed some important information, you're probably already shouting at them!

Our primitive brain is involved in processing and regulating emotions, sexual arousal, and memory. It plays an important role in the body's response to stress, and it processes the body's response to smells.

> It would be quite wrong to describe the primitive brain as negative or obsessional. Although there are times when it makes illogical choices!

So what's the problem with the primitive brain? How come we do see clients? What has gone wrong for them that hasn't for everyone else? The answer is almost always stress. Our bodies are designed to undergo some stress – we're resilient. In fact, we even go to the gym to put our bodies under stress to build up our muscles and fitness levels. But our bodies aren't designed for long periods of stress. That's when the trouble starts. And sometimes, it doesn't take much for the straw to break the camel's back and a person to find themselves suffering from anger, anxiety, or depression.

That's why our usual techniques of helping people to empty their stress bucket works so well. We are helping to reduce the amount of stress in their life. It doesn't mean that they no longer use their primitive brains and are firmly ensconced in their intellectual brain – far from it. What it does mean is that they can continue to use their primitive brain for all those low energy-consuming activities as before. It means that they can burn more energy by using their intellectual brain when they need to (like doing that maths exam, or arranging the staff conference, etc). And, importantly, it means that their primitive brain isn't trying to protect them in its own unique way.

So don't treat the primitive brain as the villain of the piece and the intellectual brain as the hero. The primitive brain is doing most of the heavy lifting for you through most of your working day. And it's doing it in a way that isn't causing anyone any problems. It's the stress in your life that makes the primitive brain perform sub-optimally. Stress is the real villain.

Unthinking thinking

Here's why we use our primitive emotional brain so much, and what the consequences can be.

Have you ever done something quite silly and then had to explain to someone why you just did that incredibly stupid thing? Do you usually find that you acted without thinking about the consequences, and everyone else is just amazed at what you've done? And yet, it's not that uncommon for people to act that way – we all do it, a lot of the time. It's just that usually it doesn't lead to such dire consequences.

The first time you ever tried to drive or ride a bike or most of the other skills you now have, you would have known very little about how things works. You wouldn't even have known what it is that you didn't know. This is what they call unconscious incompetence. After a while, you would have found out that there are all sorts of things to know about how cars work (like gears and handbrakes, etc) or bicycles, or your preferred musical instrument. This is the stage they call conscious incompetence. After a long time and a lot of hard work, you gradually got to know how to drive on a busy road (or cycle one handed, or read treble and bass clef music while playing at the same time), but you needed to think hard about what to do to get the job done correctly. This is the stage called conscious competence. After a while, with plenty of repetition, you find you can do lots of things without thinking. This is what they call unconscious competence. It's the stage you're probably at when you drive from one place to another because you don't remember very much about the drive – you've been on automatic pilot. You've been using your unconscious competence at driving. And it's this unconscious competence that can lead to so many problems!

Your brain works like this because, 99 percent of the time, everything works out perfectly and your brain uses a minimal amount of oxygen and glucose and energy to make it happen – so it feels that it has done well and tries to do the same thing again. But when something goes wrong, it goes wrong big time. That's because you're not checking what you're doing as you go along, and you plunge headlong into the world of total chaos. And that's why you end up in your bosses office trying to explain yourself – and you can't because you really didn't think too much about what you were doing, you were running on automatic.

So, how do we get to a brain that can lead us to creating such chaos in our lives? It's all down to evolution. The first part of the brain to evolve was the brain stem, which is the bit that tells your heart to beat and your lungs to breath. Then there's your limbic system (primitive brain), which is all that most mammals have to use. This gives us incredibly fast thinking and quickly accesses your habits, and uses simple rules of thumb (heuristics) to do frequent tasks. This is where you are when you're in the zone doing a sport. It's also where you are when you're on automatic pilot. It's also a bit Homer Simpson-ish in that it's selfish and childish. The third part is the cerebral cortex (intellectual brain), which is the Mr Spock part of your brain. It's logical, it solves complex problems, but it's slow – and it uses lots of energy and glucose and oxygen. These three areas all interconnect and the borders between one and the other can be a bit fuzzy!

As I say, the great thing about the cerebral cortex is that it can solve very complex problems. It's the bit of your brain you need to use at work when there are issues to resolve. It's unemotional and makes coldly logical choices. But it can be a bit of a pain if you're a sports person, or you need to learn a new skill, or in other circumstances. It's the bit of your brain that gives you the yips – that sporting problem that means you don't perform as well as you know you can. It seems that the more you think about something using the cerebral cortex, the harder it becomes to pull the habit or heuristic out of the primitive brain and use that. It's what they call paralysis by analysis. You overthink the task at hand. It's why it can be hard to teach someone else to drive because we don't usually think about what we're doing when we drive. And when we do think about it, we're not able to drive so well.

For much of the day, you want to be in your intellectual brain solving all sorts of problems, but there are times when it's the worst place to be. You can solve the yips problem by filling your intellectual brain with a mantra – often by saying the word 'smooth' if you're a golfer, or singing a song. And that lets the primitive brain do what it does best – run your habits that you've perfected over the years, such as your golf swing.

But how can you organize your life so that you stay out of your boss's office? One solution is checklists, which is what they now use in operating theatres and at major fires. This provides a simple way to indicate that each task has been completed. Better than simple checklists, are (what's called) fast and frugal trees. These help people make the best decisions in crisis situations (see Figure 1).

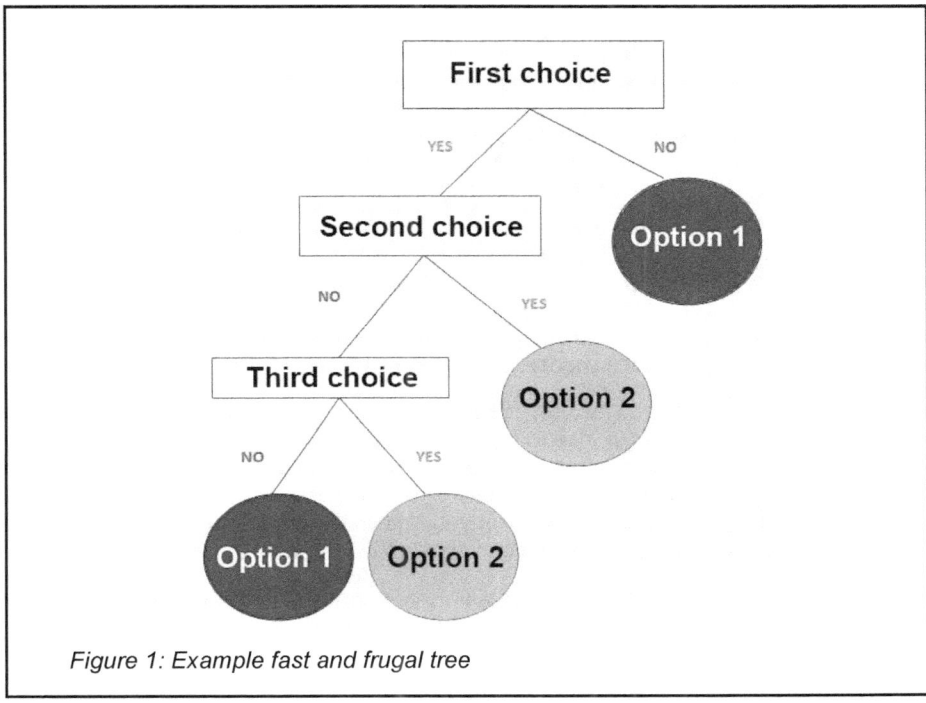

Figure 1: Example fast and frugal tree

So for speedy and efficient working, use your unconscious competence (primitive brain) to save energy and get jobs done without needing to think too much about what you're doing. And this is what most people do much of the time. But also run a check (using your intellectual brain) before you press the send button on that e-mail telling your boss what you think of him, or any other significant action.

Losing the plot

Although using your primitive brain most of the time works perfectly well, not only can it lead to incredibly stupid mistakes being made, it can also be the source of those feelings of panic – such as when you've just been asked to stand up and give a presentation to a room full of people.

The area most associated with the primitive brain, the limbic region, contains:

- The thalamus, which is where most sensory nerves pass through (except for smell).
- The hypothalamus, which is a small area that can be thought of as being full of dials that can be turned to keep everything at the right level (homeostasis). So if it's too hot, it turns the dial to make you sweat. If it's too cold, it turns the dial to make you shiver – you get the idea. The other word associated with this area is allostasis, which is where it can reset its dials to a new higher or lower level. When an animal goes into hibernation, the natural levels need to be lower. Or you're in flight or fight mode, and the levels need to be much higher.
- The amygdala (Greek for almond), which is the seat of your emotions.
- The hippocampus (Greek for sea horse), which is all to do with memories.

When a stimulus arrives in the thalamus from a sensory nerve (eg the eye), a copy goes to the intellectual brain, and a copy goes to the amygdala. If the amygdala recognizes it as a threat, it kicks off the fight or flight (or freeze) response. The hypothalamus (with all its metaphorical dials) turns everything up. It tells the pituitary gland to produce hormones, which tell the adrenal gland to produce cortisol – referred to as the HPA axis. And sympathetic nerves tell the adrenal gland to produce adrenalin and noradrenalin. Remember that the primitive brain is so much faster than the intellectual brain. So, when you've just been asked to stand up and talk to the other people in the room, you will have felt that shot of adrenalin coursing through your body long before your intellectual brain could take control and remind you that you know all about the topic and these other people are friends.

This is sometimes referred to as 'amygdala hijack'. This is where you have a strong and immediate emotional response to a situation that takes over your whole focus of attention. In more primitive times it was a way of identifying whether a shadow or movement in the trees represented something you could eat, or, more importantly in terms of survival, something that could eat you. Afterwards, you often feel that the response was in appropriate – you weren't going to be eaten by those other people!

You may remember experiencing amygdala hijack from school days. When you'd learned your tables or a poem or whatever, but the very presence of some bullying teacher standing next to you caused the amygdala to respond emotionally, and your intellectual brain was totally unable to recall what seven eights are or what happened next to Sir Patrick Spens in the ballad. And instead of giving an answer that you knew well, you sat in silence or made hesitant vocal noises.

The problem in meetings, is that when one person is experiencing amygdala hijack, it's very easy for other people to mirror them and also go into amygdala hijack. The result is a room full of people who are using the emotional part of their brain and not the intellectual part. As you can imagine (or as you may well have experienced), no sensible decisions will be reached. Often, two people are so used to arguing with each other that they almost immediately go off into amygdala hijack mode whenever they sit down in a room together. It becomes very easy to press the right buttons to get them to lose all rational control and become totally emotional.

The important thing is to get used to using your intellectual brain to get back in control. So that you appear to be more like the Fonz – always cool – or Mr Spock – always logical, rather than like Homer Simpson – out of control all the time. It's useful to be able to stay in control and not lose the plot. The more you practice it, the easier it becomes to get back control. If you need to help someone else, then get them to breathe and count to ten. An alternative is to press on the vagus nerve. You often see people naturally doing this when they are surprised by an event. They will press their hand in the centre of their chest, which presses on the vagus nerve (in a similar way to taking and holding a lungful of air) and helps calm them down.

Too fast, too furious

For many people, life can be stressful. And this can have a bad effect on your primitive brain. Let's be clear, before we move on, that some stress is good. That's what you're doing when you're training. You're putting your body under stress and, as a consequence it becomes stronger. For example, you make more muscle tissue. And some stress you can cope with using your natural resilience.

But there are times when you seem to be under a lot of stress – stress that seems to be almost continuous for a long period of time. Under these circumstances, your brain predominantly uses its limbic system rather than the cerebral cortex – and that reduces your ability to be cool and logical. You become more-and-more emotional about things. The primitive brain doesn't have any strategies for dealing with continuous stress. It's designed for dealing with short periods of stress – when there's movement in the undergrowth that might be a predator; when there's no food or warmth for a day or two; or when you're under attack.

So how does the primitive brain deal with these long stressful periods? It does what it would do in the short-term: it makes you very anxious, or it makes you depressed, or it makes you angry. For some people, they are continuously in a state of alert. They are hyper-vigilant. Their bodies are full of cortisol (a stress hormone). They tend not to be ill with colds. But if they do relax, perhaps on a family holiday, they tend to catch those colds that they've previously kept at bay. Alternatively, people start to feel depressed, and they continue to feel depressed over long periods of time. The hypothalamus changes from homeostasis (keeping everything in a constant state) to allostasis (keeping everything in a new level or state). The person's energy levels are kept running on low, and gradually everything reduces – including interest in things that previously would have been fun or attention-grabbing. Or they get angry at the slightest thing. This should be a defence response to an attack, but now everything is treated as an attack.

What can you do when your own brain is making choices that aren't right for you? How can you get out of these downward spirals? Is there a way to help yourself when you are going through this? The answer, luckily, is yes. To start with, it's important to get the key three activities right. These are sleeping, eating, and exercise. If you start to get enough sleep (and that means no more all-nighters, no more being on call and getting middle of the night phone calls), you get into a better position to deal with life. Eat properly means eating breakfast, reducing your carbohydrate input (no more sugary drinks), and no late night eating. Exercise is so good for you. If you really don't have much time, try HITs (High-Intensity Training). Exercise helps create brain cells and burns off cortisol.

The next thing is to give yourself permission to relax. Do whatever you find relaxing and enjoy the activity. Don't feel guilty because you should be at work and don't check your e-mail or social media. You can imagine a metaphorical bucket that's been filling up with stress; sleep and exercise are two of the best ways to empty that bucket. And you can start filling up your happiness bucket by doing things you enjoy and spending quality time with friends and family. What this does is give your intellectual brain a chance to cut in. Your intellectual brain is slower, but thinks logically and comes up with innovative ideas. It can start to put into perspective whatever's stressing you. It can start to come up with new ways for dealing with the situation. And it can help you act on these ideas and gradually reduce and then remove the stress. And soon you're back to your old self.

You'll find when dealing with stressed people that although you might have good ideas to help them, when they're locked into their emotional brain they can't think logically and won't accept them. And that's why you'll need to encourage them to get eating, sleeping, and exercise right. And then encourage them to relax.

You'll notice a huge improvement in their performance at work. Their problem-solving abilities will be better because they'll be using their intellectual brains, and they'll be much happier. And that's good for everyone.

The primitive brain is great for routine tasks and highly-practiced sporting activities. It's the part of the brain that helps you to catch a ball quickly and easily. But when you're put under long periods of stress, it has a very limited repertoire of techniques to deal with the situation – none of which are that successful. Many clients who come to see us are beginning to struggle with stress, and we can help them to deal with it straightaway – otherwise their primitive brain will be working too fast and too furious for a successful outcome.

Mindfulness and solution-focused hypnotherapy

How does mindfulness fit in with solution-focused hypnotherapy techniques?

Mindfulness is very popular at the moment as a way of reducing stress. Mindfulness originated as part of Buddhist practice, and was popularized by Jon Kabat-Zinn with his Mindfulness-Based Stress Reduction (MBSR) programme. It can be thought of as a mental state that can be achieved by focusing a person's awareness on the present moment, while at the same time they calmly notice and accept their feelings, thoughts, and bodily sensations without judgement.

It's not easy to do because people tend to start thinking about things, rather than focusing their awareness on, for example, just their breathing. The three key concepts of mindfulness are awareness, non-judgement, and living in the present. That means stopping the mind reflecting on the past or projecting (imagining) the future. With mindfulness, a person cultivates a positive and grateful attitude, and can let go of the need to be right.

> An ICM Survey of GPs in the UK found:
>
> 72% of GPs think that it would be helpful for their patients with mental health problems to learn mindfulness meditation skills;
>
> 66% of GPs say they would support a public information campaign to promote the potential health benefits of mindfulness meditation;
>
> 64% of GPs think that it would be helpful to receive training in mindfulness skills themselves.

Mindfulness is meant to help an individual by training their brain (neuroplasticity), improving relationships, boosting creativity, reducing depression, reducing chronic pain, giving deeper meaning to life, reducing stress and anxiety, controlling addiction, regulating eating habits, and increasing happiness.

The thing to realize about mindfulness is that is that it isn't a technique because, fundamentally, mindfulness isn't goal-oriented! Mindfulness includes a number of techniques that can be used, but, if a person practices mindfulness to achieve a goal, the mindfulness itself has less potency. It's a difficult idea, certainly. But if you use mindfulness to achieve a result, you are introducing a bias (think of it like a scientist trying to see what the result of an experiment will be) and that means you are not trying mindfulness wholeheartedly. For example, relaxation is very often a side effect of a meditation, but it shouldn't be the goal of meditation.

When you approach mindfulness, it's important to remove any 'musts', 'shoulds', and 'oughts' from your thinking because these are setting rigid rules, and mindfulness doesn't work like that. Mindfulness is simple, but not easy. It's more to do with the natural flow of things, taking you where it wants to go, rather than you setting out with a goal in mind.

The word 'mindfulness' can also be translated as 'heartfulness'. What that means is that there's more to mindfulness than just focus of attention, there's a strong emphasis on giving attention to anything that can be perceived with a sense of warmth, kindness, and friendliness, and avoiding self-criticism and blame.

> "Give up judging yourself and others; acceptance feels so much more peaceful."
> Vedam Clementi

Without mindfulness, the 'ego' takes an event and tends to develop a story or a drama. The seed of an idea develops and expands until it's larger than it really is. People feel the need to be right and this can lead to major problems. On the other hand, mindfulness brings a realistic acceptance into the picture, working with the idea that we cannot always be right and letting go of the need to be in control, develop drama, or be liked by others. Observing these tendencies of the mind and acknowledging them can be very centring and calming for a person when dealing with difficult situations.

So, let's have a look at some of the things you might do with mindfulness – some of the exercises you might want to try.

Mindful meditation:

1. Turn off all appliances, and decide how long you want to meditate for (and set a gentle-sounding alarm), dress comfortably, find a comfortable place to sit where you can be upright, and dim the lights slightly.
2. Sit as comfortably as possible and close your eyes.
3. Take three deep breaths; imagine you are breathing out all the stress and anxiety that your body contains each time you exhale.
4. Become aware of where you are holding tension in your body. When you become aware, actively release the tension and feel the muscles becoming heavier and relaxing. Pay particular attention to your jaw, shoulders, back, and pelvis.
5. Now that you have settled into your pose, take three more deep breaths in and out.
6. Now focus only on your breathing and the rhythm of the rise and fall of your chest. Bring your attention only to your breathing for around five minutes.
7. Make a mental note of how you feel in your body now.
8. Now begin to focus only on now; if you find your mind interrupts, come back to focusing on your breathing.
9. Your mind will bring in thoughts that want to take you out of now – don't judge this in any way; just notice the thought and then let it go. You are not trying to stop thinking – you are just not pursuing or indulging the thoughts.
10. Gradually, you will notice spaces between the thoughts; this is OK, too.
11. You are becoming more-and-more aware, in a non-judgemental way – just being in the moment, with no habitual thoughts or judgements.

> Focus your attention on your breath.
>
> If your attention wanders from your breath, acknowledge the current focus.
>
> Then redirect your attention back to the breath.

12. Now you are meditating. You may stay in this peaceful state for as long as you wish to. The longer you stay in this state, the deeper you will go into an altered mind state, a state where you can just be, and accept all that is there within you.
13. You may become uncomfortable if you sit for long. If this happens, simply adjust your pose slightly to release the discomfort.
14. When you are ready, or your alarm sounds, slowly allow the thoughts to return to normal. You will notice them coming back, which will highlight how different the absence of thought feels.
15. Before opening your eyes, focus again on your breathing and how different your body feels.
16. Open your eyes.

Creating internal awareness:

1. Think of a situation from the past that upset you or made you feel sad.

    Spend some time thinking about that situation and noticing how it feels in your body. This process will help you to deal with the thoughts in a mindful way.

2. Instead of expanding those thoughts and feelings, stop – begin with the intention not to continue with the usual thought process.

3. Bring your awareness from your head and emotions, to your body.

4. Start to observe how that thought or emotion feels in your body only.

5. If you can name the feeling, do so. But if you can't, just say, "This is how I'm feeling now".

> Before you speak, THINK...
>
> T – is it True?
> H – is it Helpful?
> I – is it Inspiring?
> N – is it Necessary?
> K – is it Kind?

6. Simply be or sit with this feeling, while always concentrating on not letting it translate into any analysing or thinking.

7. Keep noticing and experiencing the feeling in your body only; after a while, the feeling should start to reduce or become more bearable.

8. Bring your awareness away from the body and back into the moment.

The power of now:

1. Sit comfortably, close your eyes and breathe in and out deeply 3 or 4 times.

2. Now, intentionally, bring your attention inwards, so that all of your awareness is within you. Make a mental note of the tension points in your body. We all have them; places where there is pain, discomfort, or where the muscles are noticeably tense.

3. Once you feel focused and centred, begin to focus only on this moment as you sit there.

4. As thoughts of the past or future arise (as they will), intentionally let each one go. Keep doing this until they are very slow or not there at all.
5. Now open your eyes and, keeping this focus, do a simple five-minute task, like washing up, vacuuming, taking a shower, or going for a walk. As you do this simple thing, keep that focus only on what you are doing, nothing else.
6. While in this moment, things will happen. There is always something negative to focus on if you choose to. But just for these moments, observe only. If you feel any negative emotion, just know that you feel unhappy and why. There is no need to add any kind of judgement of yourself or anyone else to this moment.
7. Once you have finished the mindfulness task, you will notice that it had a peaceful, timeless quality and that there is less tension in your body. The secret to mindfulness is to keep coming back to the moment, so that after a while the old thought patterns and feelings in the body begin to disappear.

Dealing with emotions:

1. Notice how you feel.
2. Give the emotion a name (if you can).
3. Accept the emotion as normal (ask what prompted it), and know it will pass.
4. Spend time investigating the reaction (how strong it is, how you feel physically, has it affected your breathing).
5. Allow and release the emotion (notice and allow your thoughts; release judgements and struggles with thoughts; breathe deeply).

Noticing and changing negative thinking:

1. Sit in a comfortable chair with your back supported.
2. Close your eyes, take a few deep breaths and wait until your notice your breathing slowing down.
3. Think of a problem or worry that is currently troubling you – something that has been on your mind.
4. Now imagine what you fear could happen – a negative outcome for around one minute
5. Stop thinking! Notice how you are now feeling in your body only.
6. Connect with an understanding of what that feeling is like in your body and make a mental note to remember it.
7. Now think of the same problem – but this time imagine a positive outcome, ie what you would most like the outcome to be, for around one minute.
8. Stop! And notice how you feel now in your body.
9. What does it feel like? Now reinforce that feeling by imagining a dial going from 1

(where it is currently) to 10, and watch as it slowly goes up the dial to 10. Feel the positive feelings increase.

10 Open your eyes and note how you are feeling now.

Exercise meditation:

> What are you grateful for today?

1 Go for a walk in nature on your own.
2 Make the intention to let your thoughts go as soon as you start walking. Then notice that you are now viewing your walk and the nature around you. Take in the colours, feel the breeze, and hear the sounds.
3 Instead of thinking about what you are seeing, just be in them, almost as if you are part of it all.
4 Now focus on your breath only, and stay aware of what you can see, hear, and feel.
5 Your attention is only in this moment.
6 The more you do this, the more you begin to appreciate what you are doing. When you finish walking, notice the thought processes starting to return.

Urge surfing:

This can be used to experience the cravings/urges in a new way and to 'ride them out' until they go away. Remember that urges pass by themselves. Imagine that the urges you feel are like ocean waves that arrive, crest, and subside. They are small to begin with, get bigger, and finally break up and dissipate.

1 Practise mindfulness
2 Watch your breath. Don't alter it.
3 Notice your thoughts. Without judging them, feeding them, or fighting them, watch your thoughts, like bubbles, simply floating away. Gently bring your attention back to your breath
4 Notice the craving experience as it affects your body.
5 Focus on one area where the urge is being felt and notice what is occurring.
6 Notice the quality, position, boundaries, and intensity of the sensation.
7 Notice how these change with the in-breath and out-breath.
8 Repeat the focusing process with each part of the body involved.
9 Use a helpful mantra, if you want, like, "this too shall pass" or "I can ride out this desire" to help replace unhelpful thoughts.
10 Be curious about what occurs and notice changes over time.

And, obviously, there are many other techniques that can be used.

So, does it fit nicely with solution-focused hypnotherapy or not? Is there anything we can learn from mindfulness that will make our practice better? Like all these things, the answer is yes and no! Clearly mindfulness is not goal-oriented, whereas that's the type of hypnotherapy we do. Mindfulness is not fond of people dwelling in the past or negatively predicting the future and that's similar to our model of the right-prefrontal cortex. Much of our work helps clients to relax, and relaxation is often a side-effect of mindfulness. Mindfulness is all about accepting things the way they are, whereas we are trying to help the client make positive changes. However, that non-judgemental acceptance helps people to not fill up their bucket. Urge surfing seems like a useful technique to help, for example, smokers resist the urge to light up. It could also help phobics to surf through those urges to run away from mice/spiders/snakes when they see them. Mindfulness is all about being positive and moving away from negativity, and our sessions are all about what's been good, and expecting to hear good things next week.

> When we're depressed or anxious there's high activity in the right prefrontal cortex. Mindfulness leads to significant increase in left prefrontal cortex activation. Short-term mindfulness practice increases our happiness level.
>
> The Anterior Cingulate Cortex (ACC) is important in self-regulation and learning from past experience (resilience) to promote optimal decision making. After mindfulness training, people have higher ACC activity and show higher performance in tests of self-regulation and resisting distractors.
>
> When we're stressed, the amygdala takes control. High amygdala activity is associated with depression and anxiety disorders. Mindfulness practice shrinks the size of amygdala and increases our stress reactivity threshold.

In many ways, mindfulness is about control and being in the intellectual brain – again something we do. With mindfulness, it is often about noticing feelings, thoughts, or behaviours, and being in control enough to dissociate and just accept them as being there. I'm not sure how anyone experiencing emotional hijack would be able to cope.

Personally, I enjoy mindfulness, but I am unsure whether it has added any more tools to my solution-focused hypnotherapy toolbox, or any greater understanding about how people's minds work.

References:

Shamash Alidina; Mindfulness For Dummies; John Wiley & Sons; 978-1118868188

https://en.wikipedia.org/wiki/Mindfulness

Mindfulness Diploma Course; Centre of Excellence; http://www.coe-onlinetrainingcourses.com/

https://positivepsychologyprogram.com/benefits-of-mindfulness-practice/

Working with groups

An examination of the theories associated with working with groups and teams.

Working as a hypnotherapist, you imagine that you will spend your time working one-to-one with a client in your consulting room. But, more-and-more, hypnotherapists are going out and speaking to groups of people. After all, we have a brilliant solution-focused message to share. And now, we are beginning to work with local businesses and, in particular, schools.

Below is an example of a letter that could be sent to schools in your area. The information we have, and the kind of work we do, makes it possible for us to work with staff as well as the youngsters at the school.

Dear *headteacher*

I would like to introduce myself and offer my services, which I believe can help your students achieve better exam results.

My name is Trevor Eddolls and I am your local representative for the Association of Solution Focused Hypnotherapy (AfSFH). Solution Focused Hypnotherapists are highly-trained clinical practitioners working with both adults and children to help them manage a variety of issues, including the reduction of stress and anxiety.

The benefits of stress reduction

Schools today are constantly striving to improve standards and ensure that their young people receive an outstanding education. Expectations are high and guidelines are often changing, which can create tremendous pressure on staff and students. Managing this pressure is critical to prevent increasing levels of stress and anxiety, and reducing stress has positive results such as:

- Reduced absenteeism
- Increased productivity
- Improved focus
- Higher levels of confidence
- Overall increases in well-being.

What is Solution Focused Hypnotherapy?

Solution Focused Hypnotherapy is based on the findings of neuroscience and psychology, and incorporates the clinically-proven aspects of Solution Focused Brief Therapy (SFBT), Neuro-Linguistic Programming (NLP), and Cognitive Behavioural Therapy (CBT).

Our therapists use their neuro-scientific understanding of the workings of the brain to help adults and children alike understand why they sometimes feel sad/

depressed, scared/anxious, or upset/angry, and what they can do to take back control.

How are our therapists qualified?

Hypnotherapists with the Association for Solution Focused Hypnotherapy all hold a Hypnotherapy Practitioner Diploma (HPD) from the National Council of Hypnotherapy (NCH) and often have many other qualifications.

They are also fully insured and hold enhanced DBS checks for work with children and vulnerable adults.

What we offer

The Association for Solution Focused Hypnotherapy and its members are dedicated to the improvement of mental well-being of both adults and youngsters alike. We offer talks and workshops in schools that will help your staff and young people:

- Recognize the warning signs of excessive stress and anxiety
- Understand how and why stress affects us the way it does
- Identify and develop practical strategies to reduce stress and anxiety
- Identify and develop practical strategies to improve confidence and performance.
- Prepare for exams by using relaxation and concentration methods.

These talks and workshops can be tailored to suit all age groups, and can be delivered in a variety of formats and lengths. They are fun and interesting and are always very well received.

I do not employ any magic or trickery, but simple concentration techniques, which allow an individual to take control of their emotional state and remain focused and confident while being challenged.

If you would like to discuss this further or to arrange a meeting, please contact me on 01249443256 or at trevor@ihypno.biz.

Thank you for your time and consideration.

Usual signature and letters after your name.

Often, when working in an organization, there isn't time available to see each person individually, so you'll end up working with a group or team of people. In that case, it can be useful to understand some of the theories about how groups work.

One of the first things to recognize is that any group will have group norms. Group norms are basically the rules of conduct indicating what attitudes and behaviour might be expected or demanded in particular situations. They are shared expectations of behaviour that set up what is desirable and appropriate in a particular group. A group norm doesn't refer to what is likely to occur, but what the majority of group members think should occur. You'll find it easier to be accepted by a group if you accept their group norms.

We've all sat through meetings and noticed how some people seem to be performing specific roles within the group. The chair or secretary are obvious roles, but the other people who speak and work through the agenda seem to be performing roles as well, but in a busy meeting, it's difficult to clearly identify or name those roles. The good news is that psychologists have done that for us. Let's have a look at some of these ideas

The best known is by Dr Meredith Belbin, who defined a team role as "a tendency to behave, contribute, and interrelate with others in a particular way" and came up with nine team roles that underlie team success. Belbin's nine team roles are divided into three broad categories – action-oriented roles, people-oriented roles, and thought-oriented roles.

Action-oriented roles include the shaper (challenges the team to improve, is dynamic and usually extroverted, enjoys stimulating others, may be argumentative), the implementer (gets things done, turns the team's ideas and concepts into practical actions and plans, may be inflexible and somewhat resistant to change), and the completer-finisher (ensures projects are completed thoroughly, pays attention to the smallest of details, finds it hard to delegate).

The people-oriented roles include the coordinator (the traditional team-leader or chairman, naturally able to recognize the value of each team member, may tend to be manipulative), the team worker (provides support, flexible, diplomatic, and perceptive, tendency to be indecisive), and the investigator (innovative and curious, outgoing and extroverted, often overly optimistic).

The thought-oriented roles include the plant (creative innovator, comes up with new ideas and approaches, can be impractical at times), the monitor-evaluator (good at analysing other peoples' ideas, shrewd and objective, often perceived as detached or unemotional), and the specialist (has specialized knowledge that is needed to get the job done, prides themself on skills and abilities, preoccupation with technicalities at the expense of the bigger picture).

Kenneth Benne and Paul Sheats wrote an article entitled *Functional Roles of Group Members* in the 1940s. They came up with 26 different roles that could be played by one or more people within a group. They defined three categories of group roles: task roles, personal and social roles, and dysfunctional or individualistic roles. Task roles relate to getting the work done. They are initiator/contributor (proposes original ideas or different ways of approaching group problems or goals), information seeker (requests clarification of comments in terms of their factual adequacy), information giver (provides factual information to the group), opinion seeker (checks to make sure different perspectives are given), opinion giver (expresses his or her own opinions and beliefs), elaborator (takes other people's initial ideas and builds on them with examples, relevant facts, and data), coordinator (identifies and explains the relationships between ideas), orienter (reviews and clarifies the group's position), evaluator/critic (evaluates proposals against a predetermined or objective standard), energizer (concentrates the group's energy on forward movement), procedural technician (facilitates group discussion by taking care of logistical concerns), and recorder (acts as the secretary).

Personal and/or social roles contribute to the positive functioning of the group. They are: encourager (affirms, supports, and praises the efforts of fellow group members), harmonizer (conciliates differences between individuals), compromiser (offers to change their position for the good of the group), gatekeeper/expediter (regulates the flow of communication), observer/commentator (provides feedback to the group about how it is functioning), follower (accepts what others say and decide even though they have not contributed to the decision or expressed own thoughts).

The dysfunctional and/or individualistic roles disrupt group progress and weaken its cohesion. They are: aggressor (makes personal attacks using belittling and insulting comments), blocker (opposes every idea or opinion put forward), recognition seeker (uses meetings to draw personal attention to themselves), self-confessor (uses meetings as an avenue to disclose personal feelings and issues), disrupter/playboy/girl (treats group meetings as fun time), dominator (tries to control the conversation and dictate what people should be doing), help seeker (acts helpless, self-deprecating, and unable to contribute), special interest pleader (makes suggestions based on what others would think or feel).

Will Schutz developed FIRO-B (Fundamental Interpersonal Relations Orientation-Behaviour). The FIRO team role are: clarifier (presents issues or solutions for clarification); tension-reducer (helps move the team along by joking); individualist (not an active team player); director (pushes for action and decision-making); questioner (seeks orientation and clarification); rebel (struggles to establish a position within the group); encourager (builds the ego or status of others); listener (maintains a participatory attitude and interest nonverbally); cautioner (expresses concern about the direction of the group); initiator (suggests procedures or problems as discussion topics); energizer (urges the team toward decision-making); opinion-giver (states a belief or opinion on all problems and issues); harmonizer (reconciles opposing positions); consensus-tester (checks for agreement); task-master (tries to keep the group focused on its central purpose).

Dr Ichak Adizes developed the PAEI Model in the early 1970s. PAEI is an acronym describing four management roles that any team or organization needs to have in order to be successful. The roles are: producer, administrator, entrepreneur, and integrator.

Peter Honey suggested that there were five team roles including: the leader (ensures that the team has clear objectives and makes sure everyone is involved and committed); the challenger (questions effectiveness and presses for improvement and results); the doer (urges the team to get on with the job in hand and does practical tasks); the thinker (produces carefully considered ideas and weighs up and improves ideas from others); the supporter (eases tension and maintains team harmony).

The Margerison-McCann Team Management Profile was developed by Dr Charles Margerison and Dr Dick McCann. It's a psychometric tool (measuring things like aptitude and personality) consisting of 60 questions that explore how a person at work prefers to: relate to others, gather and use information, make decisions, and organize themselves and others.

Using the information, it would be possible to see a person's preferred role. Margerison and McCann identified eight roles, which are: reporter/adviser (enjoys giving and gathering information) creator/innovator (likes to come up with new ideas and different approaches to tasks), explorer/promoter (enjoys exploring possibilities and looking for new opportunities), assessor/developer (enjoys analysing new opportunities and making them work in practice), thruster/organizer (likes to push forward and get results), concluder/producer (likes to work in a systematic way to produce work outputs), controller/inspector (enjoys focusing on the detailed and controlling aspects of work), upholder/maintainer (likes to uphold standards and values and maintain team excellence).

Henry Mintzberg published his ten management roles in *Mintzberg on Management: Inside our Strange World of Organizations*, in 1990. The ten roles are: figurehead, leader, liaison, monitor, disseminator, spokesperson, entrepreneur, disturbance handler, resource allocator, and negotiator.

Once you're aware that there are different roles that people can play in a group, it becomes easier to nip the more destructive behaviours in the bud because you recognize them for what they are. You can also reinforce any behaviours that will benefit your work with the group. You may argue that people tend to do this automatically, but I'm not sure that people who are new to dealing with groups, and who are dealing with a group of people who are new to them, are quite so well able to modify the behaviour in the group.

As more therapists start to work with local companies, they will find themselves working with groups of people. Knowing how groups work will help us to get our message across.

References:

https://en.wikipedia.org/wiki/Team_Role_Inventories

https://www.mindtools.com/pages/article/paei-model.htm

http://www.psych-pcs.co.uk/Team_Development_Honey's_5_Team_Roles.pdf

https://www.mindtools.com/pages/article/newTMM_85.htm

https://www.cpp.com/Products/firo-b/firob_info.aspx

https://www.mindtools.com/pages/article/newTMM_58.htm

Positive psychology

What makes up positive psychology?

Positive psychology uses scientific understanding and effective intervention to aid in the achievement of a satisfactory life. Its focus is on personal growth. According to positive psychology, happiness is improved and affected in a large number of different ways, for example: social ties with a partner, family, friends and wider networks through work, clubs, or social organizations. As we suspected, happiness increases with increasing financial income, but it reaches a plateau and no additional pay rises make you any happier. It's also worth noting that physical exercise correlates with improved mental well-being.

As well as helping people change their negative style of thinking about other people, their future, and themselves, positive psychology also helps families and schools to allow children to grow; and it can be used to create workplaces that aim for satisfaction and high productivity.

Positive psychology focuses on: positive emotions (being content with your past, being happy in the present, and having hope for the future), positive individual traits (your strengths and virtues), and positive institutions (strengths to improve a community of people). One problem people often have with everyday life is remembering the good parts. We've all done it – most of the day was good except for an hour in the afternoon when things went wrong. And that's the bit we remember and tell people! One way to get an accurate record of how clients feel during a typical day is to have people (scientists) use beepers to remind them to write down the details of how they currently feel – hopefully not irritated because a beeper has just gone off! Basically, this illustrates the difference between the 'experiencing self' and the 'remembering self'. Daniel Kahneman identified a cognitive bias that he called the peak-end effect. What that means is that people remember the dramatic parts of a day and the end. So try not to let your clients leave a session with you without giving them a few minutes of a pleasant experience – particularly on a bad day, because that will colour how they remember the whole day.

Martin Seligman came up with the acronym PERMA (Positive emotions, Engagement, Relationships, Meaning and purpose, and Accomplishments) for well-being. Positive emotions include happiness, joy, excitement, satisfaction, pride, and awe. Engagement refers to involvement in activities that draw and build on a person's interests (what Mihaly Csikszentmihalyi called 'flow'). It involves passion for and concentration on the task at hand. Relationships are about receiving, sharing, and spreading positivity to others. Meaning (or purpose) drives people to continue striving for a desirable goal. Accomplishments are the pursuit of success and mastery, which may not result in positive emotions, meaning, or relationships.

You can see that PERMA can apply to all aspects of a client's life. They need to remember when they've felt 'positive emotions' about different things during the day (experiencing self). That's why asking clients, "what's been good" is such a positive technique. Clients can spend large parts of their day in the 'flow' – totally 'engaged'. It may be when they're solving a problem and they're completely absorbed in the task, or when they need to use all their knowledge and experience to identify a solution. That could be solving a Sudoku problem, a crossword, or watching a football match. And when they're chatting positively to friends and family, that's ticking the 'relationships' box. 'Meaning' is what drives us to achieve our goal, which could be to become a non-smoker because they see themselves as a non-smoker. And finally 'accomplishments' makes us study to pass exams or learn to drive or play a musical instrument better.

All these things can happen every day to our clients. And if we did measure every ten minutes how happy our clients were, we'd probably find that their experience of each day was actually quite good rather than, perhaps, how they say they remember it. And, of course, we should make sure that they leave a hypnotherapy session on a high point – whatever that might be. And, maybe, we could encourage them to run or cycle to our sessions. If we use positive psychology techniques to help make clients happier, we are also making them more confident and better able to deal with the world.

Making changes

Heraclitus of Ephesus, who lived from around 535BCE to around 475BCE, said: "The only thing that is constant is change". Heraclitus was famous for his insistence on ever-present change as being the fundamental essence of the universe! He also said: "No man ever steps in the same river twice". So if things are always changing, what's the best way to help a client make the changes that they want in their life?

Positive psychology has a technique for managing change called 'appreciative inquiry'. In fact, it's a process for implementing and achieving change, and was developed in the 1980s by David Cooperrider and Suresh Srivastva. It's a strength-based approach to managing change.

Rather than starting with negatives, appreciative inquiry starts by looking at what's working well, and builds on this information to develop a better and more successful future. Focusing on what works well generates the energy, enthusiasm, and engagement necessary to generate positivity and deal with the negative more effectively than traditional approaches. There are four stages to appreciative inquiry, and, at the start, a client has to identify a positive core. That means that the client doesn't start by looking at the problem that brought them to see you, instead they start by looking at the positive future that they want to achieve. Once this positive core (also called affirmative topic) has been identified, the client can move on to stage one.

Stage one is the 'discovery' stage, where the client looks for positive stories from friends and family about what's working well in their life. This helps flesh out the positive future. Step two is the 'dream' where the client creates a compelling and positive view of their future. It answers the question; "what might be?" It's all about taking the best of what's wanted (from stage one) and projecting it into the future. Stage three is the 'design' stage. Basically, the client starts to answer the question: "how can it be?" This helps to ground the vision of the future, and be both motivating and challenging. It's only in stage four that the practical work of turning the vision into reality starts. The theory is that appreciative inquiry creates its own momentum, and clients will spontaneously progress the change that they are passionate about. This fourth stage is called the 'destiny' stage. At this stage, it is possible for spontaneous improvements to the client's life to be allowed by building positively on what has gone before.

The lesson to take from this is that positive psychology says don't let a client look at what's wrong with their life and try to fix it, instead get them to look at what's right with their life and try to add to it.

Making the right choices

It goes without saying, doesn't it, that having a few things to choose from is good, and having lots of things to choose from must be even better? This is what democracy is all about. When I want to buy some new product, I don't want to be stuck with only one vendor. And I want to be able to choose between different features on different products. Or do I?

There have been various studies in psychology that show our well-being is increased when we feel that we can control our destiny, and that limiting personal choice reduces an individual's feelings of well-being. And that seems to make sense. But having more choices, it seems, isn't good for us. One expert, Barry Schwartz, goes so far as to say that while some choice is good, it doesn't necessarily mean that more choice is better. Schwartz even refers to this as the tyranny of choice.

In fact, Alvin Toffler, over forty years ago, theorized that, faced with too much choice (he called it 'overchoice') in too short a period of time, we would find decisions harder to make and they would take longer to make. He suggested that, ultimately, too much choice would lead to depression, distress, and neurosis! Having too much choice causes people to worry, and is likely to lead to lower levels of well-being.

Psychologists reckon that too much choice can have a demotivating effect. Choice overload can hinder important decision-making, especially where there are costs associated with making the 'wrong' choice, and where it takes the chooser a significant amount of time and effort to make the informed decision. Clients can feel that when they're trying to choose which is the best therapy to help them. Each comes with glowing reviews from people who've already tried it, and each has a cost per session and an unknown number of sessions. There's also an issue over who to trust with the evaluation. Some therapies are based on proven science (eg neuroscience and psychology) and some are based on 2000 years of tradition. And some claim to work by saying that science doesn't know everything. It can be very hard for the non-expert client to make a choice.

Herbert Simon divided people into 'maximizers' and 'satisficers'. Maximizers are people who make a study of all the available options before, finally, making their decision. On the other hand, satisficers will make a good enough decision – so they look at options until they find one that meets their requirements (and then stop). While being a maximizer seems like the only thorough approach and the one that we should all be taking, there are downsides. For example, it takes a considerable amount of time and effort to look at all the options available. This time and effort could be spent in other aspects of your life. The other issue is that, at the end of the day, the maximizer might still not be sure in their own mind that they really have made the best choice possible. Apparently, maximizers have higher expectations from their choice, which can lead to disappointment when they get it. Plus they are always worried that they may have missed a bigger/better/faster one, or that a better one may have been announced while they were looking at the existing choices. Lastly, they tend to blame themselves, if the product/therapy doesn't perform as well as they hoped – often thinking that they should have spent longer looking at the options available.

Bear this in mind when giving clients choices – if you ever do. Remember that some choice is good but too much is very difficult to handle.

References:

Bridget Grenville-Cleave; Positive Psychology – A Toolkit for Happiness, Purpose and Well-being; ISBN: 9781848319561

https://en.wikipedia.org/wiki/Positive_psychology

What can we learn from Sufism?

Let's investigate what mystical Sufis can teach us about dealing with the world.

Sufism is one of those belief systems that has much in common with our way of looking at the world. Let's find out more about who they are and what they say.

All Sufis are Moslems, but not all Moslems are Sufis. Sufis are more concerned with God than other Moslems. According to Ibn Khaldunin in the 14th Century, Sufism is, "dedication to worship, total dedication to Allah most High, disregard for the finery and ornament of the world, abstinence from the pleasure, wealth, and prestige sought by most men, and retiring from others to worship alone." Sufism started in the very early days of the Islamic religion.

Let's have a look at some Sufi proverbs:

- Abundance can be had simply by knowingly receiving what has already been given.
- Asking good questions is half of learning
- Behind every adversity lies a hidden possibility.
- Grasp the moment; you can't power a mill with water that has already passed by.
- There is a difference between spending a night with a lover and a night with a toothache.
- There would be no such a thing as counterfeit gold if there weren't real gold somewhere.
- Those that have time and search for a better time will lose time.
- When a pickpocket sees a saint, all he sees are his pockets.
- Wise company can also make you wise.

That's not too different from what we might say.

Or here are some quotes

- You yourself are your own obstacle – rise above yourself (Hafiz)
- I caught the joy virus last night when I was out singing beneath the stars. (Hafiz)
- All wisdom can be expressed in two phrases: what is done for you – allow it to be done; what you must do yourself – make sure you do it. (Khawwas)
- A person cannot live in 'if'. (Maulana) Jalal Al-Din (or Jalaluddin) Rumi
- You already possess the powerful mixture that will make you well – use it. (Maulana) Jalal Al-Din (or Jalaluddin) Rumi
- Appear as you are; be as you appear. (Maulana) Jalal Al-Din (or Jalaluddin) Rumi

- Your mission isn't to look for love, but simply to search and locate the barriers within you that have formed against it. (Maulana) Jalal Al-Din (or Jalaluddin) Rumi
- We waste our energy designing and carrying out plans to become what we already are. (Maulana) Jalal Al-Din (or Jalaluddin) Rumi
- Become your own fortune… Seek the bounty within yourself. (Maulana) Jalal Al-Din (or Jalaluddin) Rumi
- All darkness is followed by sunshine. (Maulana) Jalal Al-Din (or Jalaluddin) Rumi
- No one should allow his mind to be a vehicle for others to use; he who does not direct his own mind lacks mastery. Hazrat Inayat Khan
- The secret of life is balance, and the absence of balance is life's destruction. Hazrat Inayat Khan
- Worrying about the faults of others is an unnecessary addition to the worry we have over our own faults. Hazrat Inayat Khan
- While man judges another from his own moral standpoint, the wise man looks also at the point of view of another. Hazrat Inayat Khan
- Do not take the example of another as an excuse for your own wrongdoing. Hazrat Inayat Khan
- It is seldom that too little is said and too much is done, but often the contrary. Hazrat Inayat Khan
- Wisdom is not in words; it is in understanding. Hazrat Inayat Khan
- Speaking wisdom is much easier than living it. Hazrat Inayat Khan
- Life is an opportunity, and it is a great pity if man realizes this when it is too late. Hazrat Inayat Khan
- The more a man explores himself, the more power he finds within. Hazrat Inayat Khan

And here are some Sufi metaphorical stories…

A bird's advice

There was a man, who caught a bird, and the bird said to him: "Release me, and I will give you three valuable pieces of advice. I will give you the first when you let me go, the second when I fly up to that branch, and the third when I fly up to the top of the tree."

The man agreed to this, and let the bird go. The bird said, "do not torment yourself with excessive regret for mistakes."

Next, the bird flew up to a branch and said, "do not believe anything that goes against common sense, unless you have first-hand proof."

Finally, the bird flew up to the top of the tree and said, "you idiot. I have two huge jewels inside me. If you'd killed me instead of letting me go, you could now have them."

"Oh no!" said the man. "How can I have been so stupid? I'll never get over this. Bird, will you at least give me the third piece of advice as a consolation?"

The bird replied: "I've been kidding you. And yet you are asking for more advice, even though you have already disregarded the first two pieces of advice that I gave you. I told you not to torment yourself with excessive regret for mistakes, and I told you not to believe things that go again common sense unless you have some kind of first-hand proof. And here you are tormenting yourself with excessive regret for letting me go, and believing that there really are two jewels inside a tiny bird like me! So here is your third piece of advice: if you are not using what you know, why are you so intent on seeking what you do not know?"

Cold day

It was a cold wintery day, and a heavily dressed man noticed Nasrudin outside wearing very little clothing. "Mulla", the man said, "tell me, how is it that I am wearing all these clothes and I still feel a little cold, whereas you are barely wearing anything yet seem unaffected by the weather?"

"Well", replied Nasrudin, "I don't have any more clothes, so I can't afford to feel the cold, whereas you have plenty of clothes, and so you can feel as cold as you want."

The crowded home

Nasrudin was talking to his neighbour, when the neighbour complained, "I'm really having trouble fitting all my family in our small house. There's me, my wife, my three children, and my mother-in-law. We all share the same cottage. Mulla Nasrudin, you are a wise man, do you have any advice for me?"

"Yes", replied Nasrudin. "Do you have any chickens in your garden?"

"I have ten", replied the man.

"Put them in the house", said Nasrudin.

"But Mulla," the man observed, "our house is already cramped as it is".

"Just try it", replied Nasrudin.

The man, desperate to find a solution to his problems with space indoors, followed Nasrudin's advice. Next day, he paid him another visit.

"Mulla", he said, "things are even worse now. With the chickens in the house, we are even more pressed for space."

"Now take that donkey of yours", replied Nasrudin, "and put it in the house too".

The man moaned and objected, but Nasrudin convinced him to do it.

The next day, the man, now looking more distressed than ever, came up to Nasrudin and said, "Now my home is even more crowded! Between my family, the chickens, and that donkey of mine, there is virtually no room to move."

"Well then", said Nasrudin, "do you have any other animals in your garden?"

"Yes", the man replied, "we have a goat".

"Great", said Nasrudin. "Put the goat in your house too."

The man once again made a fuss and seemed anything but eager to follow Nasrudin's advice, but Nasrudin once again convinced him to put yet another animal in the house.

The next day, the man, very angry, came up to Nasrudin and exclaimed, "my family is really upset now. Everyone is at my throat complaining about the lack of space. Your plan is making us miserable."

"Fantastic", Nasrudin replied, "now take all of the animals back outside".

So the man followed his advice, and the next day, he dropped by and said, "mulla, your plan has worked like a charm. With all the animals out, my house is so spacious that none of us can help but be pleased and uncomplaining."

The best light

A man noticed Nasrudin staring intently at the ground outside his door.

"Mulla", he said, "what are you looking for?"

"I'm looking for a ring I dropped", Nasrudin replied.

"Oh", the man replied as he also started searching. "Where exactly were you standing when you dropped it?"

"In my bedroom", Nasrudin replied, "not more than a foot in front of my bed."

"Your bedroom?!" the man asked, amazed. "Then why are you searching for it out here near your doorway."

"Because", Nasrudin explained, "the light is much better out here".

What's right under your nose?

Nasrudin, the smuggler, was leading a donkey that had bundles of straw on its back. An experienced border inspector spotted Nasrudin coming to his border.

"Halt", the inspector said. "What is your business here?"

"I am an honest smuggler!" replied Nasrudin.

"Oh, really?" said the inspector. "Well, let me search those straw bundles. If I find something in them, you are required to pay a border fee!"

"Do as you wish", Nasrudin replied, "but you will not find anything in those bundles".

The inspector searched intensely and took the bundles apart, but couldn't find a single thing in them. He turned to Nasrudin and said, "I suppose you have managed to get one by me today. You may pass the border."

Nasrudin crossed the border with his donkey while the annoyed inspector looked on. And then the very next day, Nasrudin once again came to the border with a straw-carrying donkey. The inspector saw Nasrudin coming and thought, "I'll get him for sure this time".

He checked the bundles of straw again, and then searched through Nasrudin's clothing, and even went through the donkey's harness. But yet again he found nothing and had to let Nasrudin pass.

This same thing happened every day for several years, and every day Nasrudin wore more and more extravagant clothing and jewellery – indicating that he was getting wealthier. Eventually, the inspector retired from his job, but even in retirement he still wondered about the man with the straw-carrying donkey. "I should have checked that donkey's mouth more extensively," he thought to himself. "Or, perhaps, he hid something in the donkey's rectum."

Then, one day, the man spotted Nasrudin's face in a crowd. "Hey", the inspector said, "I know you! You're that man who came to my border every day for all those years with a donkey carrying straw. I must talk to you." Nasrudin came towards him and the ex-inspector continued talking. "My friend, I always wondered what you were smuggling past my border every day. Just between you and me, you must tell me. I must know. What in the world were you smuggling for all those years? I must know!"

Nasrudin looked at the man and said simply, "donkeys".

You can't please everyone

Nasrudin and his son were travelling with their donkey. Nasrudin preferred to walk while his son rode the donkey. But when they passed a group of bystanders, one scoffed, "Look! That selfish boy is riding on a donkey while his poor old father is forced to walk alongside. That is so disrespectful. What a horrible and spoiled child!"

Nasrudin and his son felt embarrassed, so they swapped round. This time Nasrudin rode the donkey while his son walked. Soon they passed another group of people. "Oh, that's detestable!" one of them exclaimed. "That poor young boy has to walk while his abusive father rides the donkey! That horrible man should be ashamed of himself for the way he's treating his son. What a heartless parent!"

Nasrudin was upset to hear this. He wanted to avoid anybody else's scorn, so he decided to have both himself and his son ride the donkey at the same time. As they both rode, they passed another group of people. "That man and his son are so cruel", one bystander said. "Just look at how they are forcing that poor donkey to bear the weight of two people. They should be put in prison for their despicable act. What scoundrels!"

Nasrudin heard this and told his son, "I imagine the only way we can avoid the derisive comments of others is to both walk".

"I suppose you are right", the son replied.

So they got off the donkey and continued on foot. But as they passed another group of people, they heard them laughing. "Ha, ha, ha", the group jeered. "Look at those two fools. They are so stupid that both of them are walking under this scorching hot sun and neither of them is riding the donkey! What idiots!"

I like some of the quotes, and I've heard the stories told as metaphors for therapy. There's plenty for modern-day hypnotherapists to learn from Sufism.

References:

https://en.wikipedia.org/wiki/Sufism

http://www.rodneyohebsion.com/sufism.htm

http://www.rodneyohebsion.com/sufi-folktales.htm

About the author

Trevor Eddolls BA, Cert Ed, MOS MI, DHP, HPD, SFBT Sup (Hyp), CBT (Hyp), Dip NLP, Dip Mindfulness, AfSFH (Exec), CNHC Registered, UKCHO Registered is a clinical hypnotherapist and psychotherapist. He's clinical director at iTech-Ed Hypnotherapy and Head of IT on the AfSFH (Association for Solution-Focused Hypnotherapy) Executive. Trevor is a Hypnotherapy Master Practitioner, a Solution-Focused Hypnotherapy Supervisor, and an NLP Master Practitioner. He has a qualification in CBT and a diploma in Mindfulness. He is a qualified Life Coach, and has a diploma in nutrition.

Solution-focused hypnotherapy, as its name suggests, focuses a client's attention on the solution to their problems rather than the causes. Evidence suggests that dwelling on what led to a problem can increase the client's issues, whereas focusing on solutions can dramatically reduce those issues.

Trevor has been seeing clients and writing about hypnotherapy, CBT (Cognitive Behavioural Therapy), NLP (Neuro-Linguistic Programming), and Mindfulness techniques for a number of years.

Before training as a hypnotherapist, Trevor worked with mainframes. He also spent many years writing books and articles, and editing well-respected technical journals about mainframe technology.

You can contact Trevor at iTech-Ed Hypnotherapy in the Wiltshire town of Chippenham.

His Web site is at www.ihypno.biz.

Facebook: facebook.com/iHypno2004

Twitter: twitter.com/iHypno2004

Instagram: instagram.com/ihypno2004

www.ingramcontent.com/pod-product-compliance
Lightning Source LLC
Chambersburg PA
CBHW072224170526
45158CB00002BA/735